YF-22
ED
047

F-22 RAPTOR

Bill Sweetman

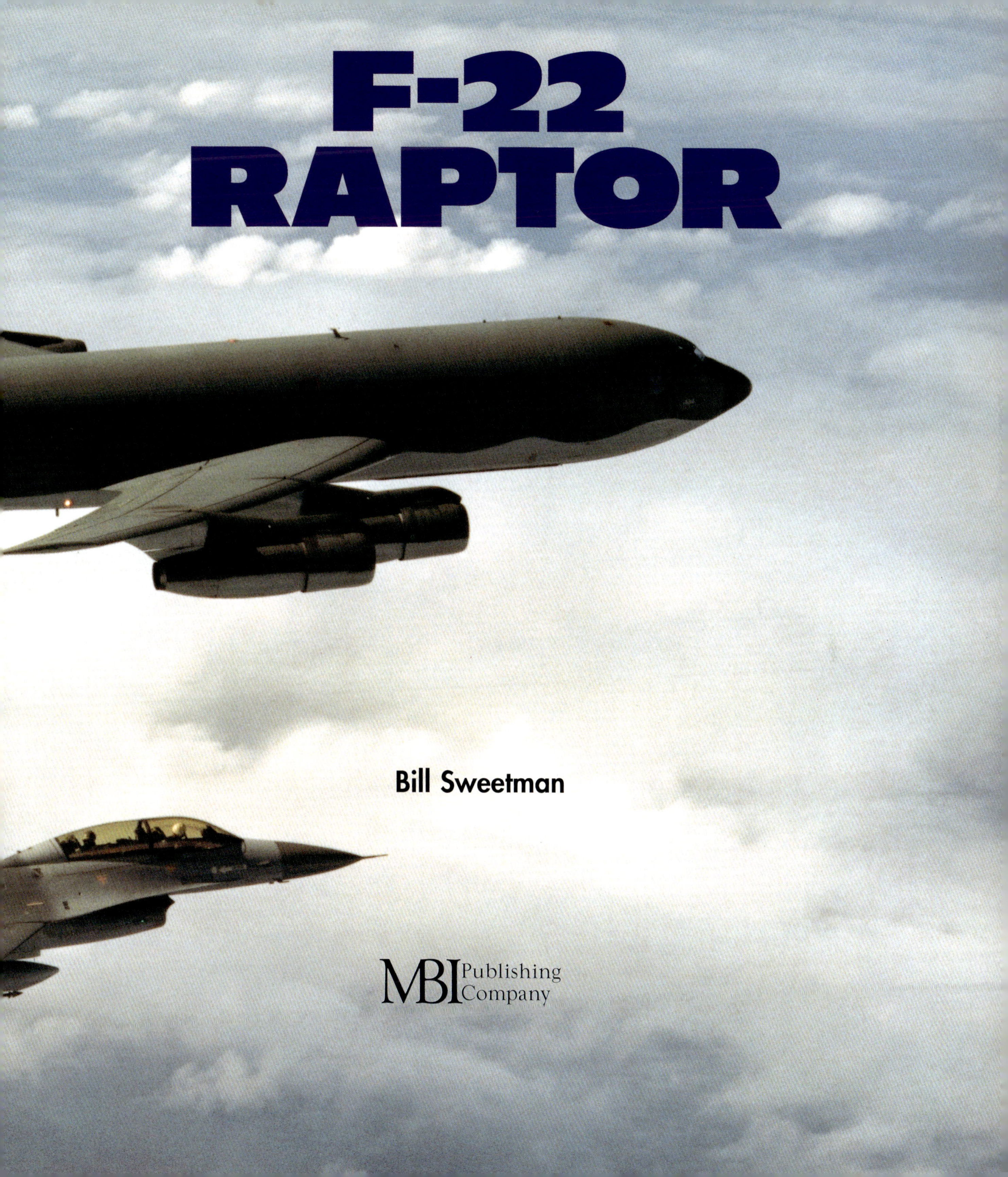

MBI Publishing Company

First published in 1998 by MBI Publishing Company, 729 Prospect Avenue, PO Box 1, Osceola, WI 54020-0001 USA.

Library of Congress Cataloging-in-Publication Data Available

ISBN 0-7603-0678-8

On the frontispiece: A Lockheed Martin engineer uses an optical-electronic "step-gap measurement tool" to check the fit of an access panel of the F-22. Panels and doors must fit within certain tolerances to maintain the fighter's stealth characteristics, and physical measurement of the entire panel is extremely time-consuming. A similar system will be deployed with operational fighters for use in the field. *Lockheed Martin*

On the title page: In-flight refueling was accomplished early in the YF-22 flight-test program and was used routinely to extend the length and productivity of each flight. The rotating in-flight refueling (IFR) receptacle is similar to that used on the F-22A. The prototypes were designed to represent the performance and handling of the final design, with similar thrust and loaded weights, but they were built quickly, sturdily, and somewhat overweight, and could not accommodate as much fuel as the production aircraft. *Lockheed Martin*

Printed in Hong Kong through World Print, Ltd.

Contents

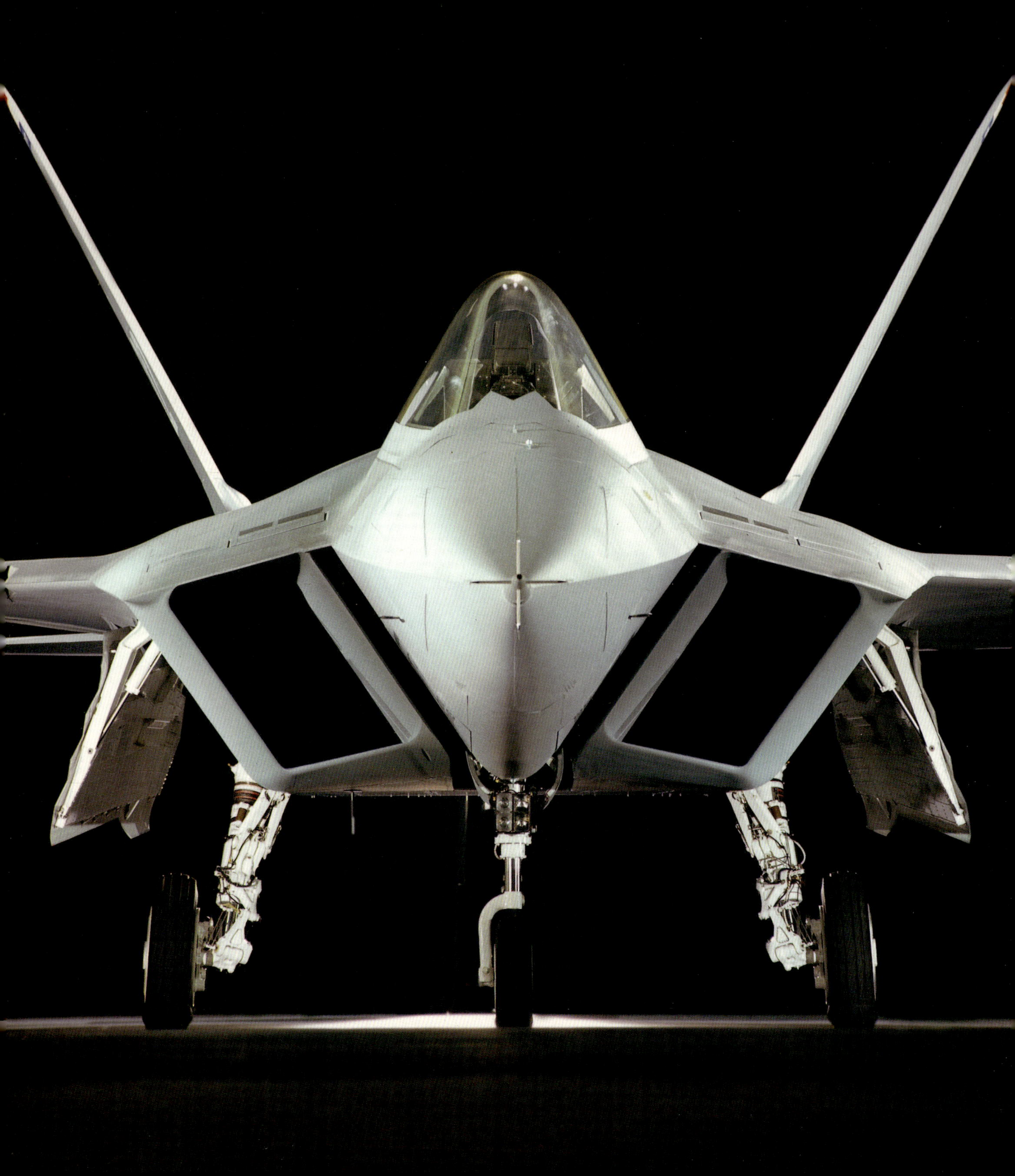

Chapter One

Hatching the Raptor

The Lockheed Martin F-22 Raptor is the greatest one-generation advance in fighter-aircraft capability in 50 years. It brings the largest increase in sustained speed since the advent of the jet, flying most of its mission at speeds that other fighters attain only in short sprints, and accelerating and maneuvering at speeds at which today's fighters struggle to maintain in a straight line. It will equal and probably surpass the agility of any other fighter. It is stealthy in all aspects against most radars and other sensors. Its integrated avionics and sensor-fused displays are a generation ahead of anything under test elsewhere.

The F-22's primary mission, air superiority, is rooted in air warfare. Air combat units are mobile by nature and can direct fire against any target within their range. The odds are in favor of the side that has the faster aircraft with greater range, operating under the fewest artificial constraints. The Raptor commander can focus his force in any area he chooses, while the defender can only respond with aircraft that can reach the battle in time.

Air superiority has been defined by the USAF as "the ability to achieve local air control at a time and place of our own choosing." Once this capability has been established, the commanders can use it to clear the way for their air strikes to break up hostile attacks long before they reach their targets, and to force hostile aircraft to act defensively, even in their own airspace. The Raptor's secondary goal is to destroy as many of the enemy's front-line fighters as possible for the smallest possible number of losses.

Eclipsed by nuclear strike and air defense missions in the 1950s, air superiority re-emerged as a critical mission in Vietnam, with the McDonnell F-4 Phantom as the first air superiority fighter of the modern era. This was not what the F-4 was designed to do. Conceived as an attack aircraft for the U.S. Navy, it was revised as an interceptor before its first flight and modified as a multirole fighter before it was forced on the U.S. Air Force as a stop-gap.

When the YF-22 was unveiled in August 1991, its resemblance to an inverted F-117 was clear. Better modeling and testing techniques allowed the designers to incorporate some curvature in the surfaces of the F-22, avoiding the totally faceted shape of the F-117. The edges between facets were rounded off, and wing surfaces were curved. *Lockheed Martin*

Before the ATF competition even started, the Lockheed Skunk Works studied fighter designs such as this CL-1980. Clearly a large aircraft with a family resemblance to the SR-71 Blackbird, it was designed to use high speed and altitude to overfly all but the largest SAMs. Vectored thrust made it reasonably maneuverable, and flip-out canards reduced its takeoff and landing speeds. *Lockheed*

As the Vietnam War continued, the Air Force wrote a requirement for its first pure air-superiority fighter since the F-100A Super Sabre. It was little wonder that it emerged in essence as a vastly improved F-4, or that McDonnell Douglas should win the contract to build what became the F-15 Eagle.

But as the first F-15 squadrons became operational in 1978, alarming evidence suggested that the new fighter's superiority might be transitory. Reconnaissance satellites passing over the Soviet flight test center at Zhukhovsky airfield, north of Moscow, discovered new fighters under test. One was a new agile fighter—the Mikoyan MiG-29, which entered limited service in 1983 and became fully operational in 1986. The other was more of a surprise: It was bigger than the F-15, and much bigger than previous Soviet tactical fighters. It was the Sukhoi T-10 prototype, ancestor of the Su-27 Flanker.

Western analysts were shocked by the speed with which the Soviet Union modernized its fighter force in the 1970s, building as many as 500 new MiGs each year. If this pattern repeated itself,

Northrop started its advanced fighter studies at the opposite end of the scale from Lockheed, with an emphasis on small, low-cost, agile aircraft. This design, the ND-102, was jointly devised by Northrop and Germany's Dornier company as a replacement for Luftwaffe F-4Fs. It had a clipped delta wing, no horizontal tail, and thrust vectoring for agility. *Northrop*

Developed under a highly secret program, flown between 1982 and 1985, and not disclosed until 1996, the Northrop Tacit Blue—nicknamed the Whale—was an important part of the ATF story. It was the first stealth aircraft to use radar, validated Northrop's all-aspect stealth technology, and proved that an unstable V-tailed aircraft could fly. *Northrop*

the MiG-29 would enter service as a one-for-one replacement for the MiG-21, with the Su-27 replacing the MiG-23. The prospect was described by one analyst as "a high/low mix, with their 'low' equivalent to our 'high.'"

Surface-to-air missiles (SAMs) presented an equally dangerous threat. The Soviet Union built many more SAMs than the West and introduced a new generation of weapons every 10 years. By the early 1980s, the Warsaw Pact had installed so many SAMs in Eastern Europe that they constituted a barrier. The barrier could be damaged by defense-suppression aircraft or overwhelmed by a large strike package, but nobody pretended that SAMs would not destroy many NATO aircraft.

This was the background to the studies of follow-on fighter aircraft that started in the late 1970s. In mid-1981, the Air Force formally identified a need for a follow-on fighter, and in November, the USAF Aeronautical Systems Division (ASD) issued a request for information on what was called, for the first time, the Advanced Tactical Fighter (ATF).

Industry responded to the request for information (RFI) with concepts that ranged from a 25,000-pound fighter (smaller than the F-16) to a Mach 3, 120,000-pound "battlecruiser." There was some common ground, summed up by a McDonnell Douglas paper that emphasized "STOL, stealth, and supercruise." STOL, or short takeoff and landing, was desired because runways would be prime targets. STOL would make it easier to repair enough runway to resume operations.

Stealth was controversial. The RFI was issued four months after the first Lockheed F-117 stealth strike fighter flew from the Air Force's secret flight-test base at Groom Lake, Nevada. Stealth clearly had a great deal of potential, both against radar-guided SAMs and in radar-missile duels with hostile fighters. A Northrop report, published in 1981 showed how a stealth fighter could see its opponent first and fire first, even if the adversary had a more powerful radar and a longer-range missile. But few people were cleared into the F-117 program, and stealth was widely regarded as an exotic technology, inapplicable to a tactical fighter.

Supercruise, or the ability to exceed supersonic speed without using afterburners, would give the pilot the option of engaging or evading opponents and put the fighter beyond the reach of smaller SAMs. Engine improvements made it look attainable.

The Air Force added some economic considerations to these operational requirements. From the P-51 to the F-15, each new generation of fighter cost more than the one before and was bought in smaller numbers. Norman Augustine, then president of Martin-Marietta (and later the architect of the Lockheed Martin merger), extrapolated the trend to the year 2054 and concluded that, in that year, the entire U.S. defense budget would buy one tactical fighter. Augustine said the Navy and Air Force could use it on alternate days.

Northrop's stealth design philosophy, combining rigorous planform alignment with smoothly curved surfaces, was instrumental in winning the B-2 Advanced Technology Bomber for the company. It put Northrop in a position to compete in ATF and forced Lockheed—shocked by its defeat—to look for new business, a step that led to the F-22. *Northrop*

In October 1982, representatives from most of the fighter manufacturers, together with planners

Boeing earned fourth place in the ATF Dem/Val contest with this V-tail, diamond-wing design. The single shark-mouthed inlet fed both engines—a unique and not necessarily desirable feature in a fighter, because of the risk that an engine failure or stall in one engine would affect the airflow into another. The inlet reappeared on Boeing's single-engine Joint Strike Fighter design. Like the F-22, it had a ventral bay for medium-range missiles and side bays for AIM-9s. *Lockheed Martin*

and requirements specialists from the Air Force, met at an aerospace engineering conference in Anaheim, California, for a half-day session on the ATF. (It was the last unclassified meeting on ATF for many years.) The outlines of today's fighter began to emerge: a supersonic-cruise aircraft, with a combat radius of 700 to 900 miles (a 20 to 60 percent increase over the F-15), and with reduced observables if possible. It would be able to take off and land on a 2,000-foot runway, and it would be easier to maintain in the field than the F-15.

Air Force planners concluded that range and supersonic persistence set a lower bound to the fighter's size, and price set a top limit. The requirements balanced at a normal takeoff weight of about 80,000 pounds for an interdiction/strike aircraft and 60,000 pounds for an air-superiority fighter.

Runner-up to Lockheed and Northrop was General Dynamics, with a tailless delta design that reflected the success of the arrow-winged F-16XL. The dramatically serrated trailing edge has the same alignments as the leading edge of the wing. GD proposed to put dual radar arrays in the leading-edge root extensions, above the inlets, and an infrared sensor in the nose. The single vertical tail, however, detracted from its all-aspect stealth. *Lockheed Martin*

Lockheed's original ATF design echoed the F-117, with its arrowhead shape and a blended forward fuselage, which might have restricted the pilot's downward view. It featured more conventional trapezoidal wings and vectored thrust to meet maneuver requirements and a horizontal tail, however. The four tails and boom-mounted horizontals survived into the final design, but little else of the plantorm did. *Lockheed Martin*

The lower body shape of the original Lockheed design foreshadowed the F-22—the overwing, gridded inlets of the F-117 would not work on a supersonic aircraft. Internally, it was completely different from the F-22, with a single weapons bay in the midfuselage housing a rotary launcher. The vertical tails are highly swept and would have been quite heavy. *Lockheed Martin*

Another part of the ATF puzzle fell into place during 1983. Both General Dynamics and McDonnell Douglas showed that the F-16 and F-15 could be modified to perform all-weather, low-level strike missions. The Air Force decided to buy a strike version of the F-15 or F-16 first (McDonnell Douglas won this contest with the F-15E), and follow that with a fleet of 750 dedicated air-superiority ATFs to replace the F-15, starting in the mid-1990s.

In 1983, an ATF System Program Office (SPO) was formed at Wright-Patterson Air Force Base in Ohio. Headed by Colonel Albert C. Piccirillo—a fighter pilot with F-4 experience in Vietnam and time in F-15s—its task was to produce a specification that met all the users' essential requirements and as many of their desires as practical. The traps to avoid were requirements of marginal value and disproportionately high cost. Every pound of equipment added to the aircraft meant a 5-pound increase in gross weight to meet the same perfor-

The F-117 gave Lockheed a strong position in stealth technology. The fundamental principle was the recognition that the largest radar target is a flat surface at 90 degrees to the radar beam, and the smallest target is that same surface, when it is rotated away from the beam in two dimensions. Because most of the radars that the designer is worried about lie close to the plane of the aircraft when it is in level flight, and the most dangerous of these are the radars ahead of the airplane, the F-117 was designed with surfaces that are swept back and canted inward. *Lockheed Martin*

mance requirements. "Early in this stage," Piccirillo remarked in an interview, "we found four or five significant drivers." These were specific requirements that added a great deal to the fighter's weight, "and making just one of them cost us 10,000 pounds." In other cases, Piccirillo said, "backing off by half a percent was important."

Because the ATF would be optimized for air superiority, the Air Force knew the ATF's approximate size and speed, and consequently, how big its engines should be. The request for proposals for the ATF engine, then known as the Joint Advanced Fighter Engine (JAFE), was issued in May 1983, and General Electric and Pratt & Whitney were awarded contracts to build and test competing engine designs in September.

At the same time, the Air Force issued concept definition contracts to the seven U.S. fighter design teams: Boeing, General Dynamics, Grumman, Lockheed, McDonnell Douglas, Northrop, and Rockwell.

The Air Force wrote a near-final requirement at the end of 1984. It called for ATF to have an operational radius of about 800 miles, enough to

After the contract award, the first YF120-powered YF-22 did not fly again, but was used as an engineering fixture at Marietta. In 1997, it underwent a cosmetic restoration and was displayed at the U.S. Air Force 50th anniversary air show at Nellis Air Force Base, Las Vegas—confusingly, with Pratt & Whitney labels on the engine access panels. This view at Nellis shows the bi-fold weapon bay doors. To load missiles, one set of bay doors is opened at a time, and the missiles are loaded from the opposite side of the aircraft. The dark patches inside the inlet are fine perforations for boundary layer control. *Bill Sweetman*

allow it to cover the entire Central Region of Europe from bases in central England. ATF would be able to cruise at Mach 1.4–1.5 throughout that segment of its mission that crossed hostile territory (up to 300 miles in and out), and it would do it at supersonic speed. It would operate from less than 2,000 feet of runway. The target for normal takeoff weight was 50,000 pounds, about the same as an F-15C with its centerline fuel tank and eight air-to-air missiles (AAM). Although, Piccirillo would say much later, the SPO fully expected a 60,000-pound airplane to emerge. The "flyaway" cost—the price of one fully equipped aircraft, averaged across the entire production run—was not to exceed $40 million in 1985 dollars.

In September 1985, the Air Force issued a request for proposals (RFP) on the first stage of the ATF program. The requirements had not changed

Because stealth requires all apertures to be closed securely and tightly, the YF-22 designers wanted no more doors than necessary—hence this very large single door for the forward-retracting main landing gear. The F-22A has a shorter, outward-retracting gear. *Bill Sweetman*

apart from a Pentagon-mandated cut in the flyaway price to $35 million. One new feature was the way in which the competition was to be conducted. The Air Force did not want to pick a winner on the basis of a paper proposal, but was not sure that it wanted a full-scale fly-off competition. Instead, the Air Force conducted a new kind of competition: a "demonstration and validation" (Dem/Val) program in which the riskiest technologies would be tested at large scale.

All seven design teams responded to the RFP. The logical favorites were McDonnell Douglas and General Dynamics, the two companies that built all the fighters that the Air Force had bought for its own use since the early 1960s. Grumman worked almost exclusively for the Navy. Rockwell had not flown a new fighter since 1956 and was extremely busy with the B-1. Lockheed had built only one operational supersonic fighter, the F-104, and the Air Force had not liked it. Northrop was totally committed to the B-2 stealth bomber and was known for small, light fighters, the kind the Air Force detested. Boeing had never built a jet fighter or a manned supersonic aircraft.

The F-117 had experienced a lack of directional stability on its first flight, and Lockheed was not about to make the same mistake with the YF-22, which emerged with barn-door-sized verticals. These constituted overkill, and the EMD aircraft's vertical tails are 30 percent smaller. *Lockheed Martin*

The conventional wisdom was wrong, but in 1984 Lockheed itself did not see ATF as a prime opportunity. Even after Northrop trounced Lockheed in the competition to build the B-2, in 1981, Lockheed focused on Navy requirements. Lockheed and Northrop each told the Navy that it was possible to build a stealth replacement for the A-6: The Navy agreed and launched the secret and ambitious Advanced Tactical Aircraft (ATA) program.

Lockheed displeased Navy Secretary John Lehman by successfully lobbying to keep the P-3

In April 1992, the second YF-22 entered a series of oscillations during a low, slow pass along the runway at Edwards, crashed, slid 8,000 feet along the runway, and burned. The pilot survived, testimony to the prototype's overbuilt structure. The aircraft could not be economically restored to flying condition, but was repaired externally and moved to Rome Air Development Center at Griffiss Air Force Base in New York, where it was mounted on a pedestal and used to test antenna designs for the F-22A. *U.S. Air Force*

production line open when the Navy tried to close it in 1982. When the Navy formed teams for the ATA contest, Northrop teamed with Grumman, General Dynamics teamed with McDonnell Douglas, and Lockheed and Boeing teamed with LTV, which had less recent carrier experience than Grumman or McDonnell Douglas. In November 1984, Northrop/Grumman and General Dynamics/McDonnell Douglas were selected to continue with the ATA program and Lockheed was eliminated. (General Dynamics and McDonnell Douglas later won the contract to build the A-12 Avenger II, but it was scrapped in 1991.)

Lockheed turned its attention to the ATF. At first, the company was an outsider, and its stealth-oriented design was (some Lockheed people felt) regarded as a "warmed-over F-117." A series of developments in 1985 and 1986 improved Lockheed's chances, however.

The most important was that the Air Force increased its emphasis on stealth. While the details of the ATF requirement were being worked out,

USAF
91480
YF-23
YF-22

Montagu and Capulet in formation: a rare image of the first YF-22 and the second Northrop YF-23 above the Mojave Desert. The two fighters were similar in overall dimensions and weight, but far apart in shape: The Northrop fighter's separate forebody and nacelles contrast with the more solid-looking YF-22. *U.S. Air Force*

the F-117 completed its flight tests. Northrop flew its Tacit Blue stealth prototype and completed most of an extensive risk-reduction program on the B-2. The key technologies behind stealth—shaping, materials, and measurement and testing—matured quickly, and Northrop proved that a stealthy aircraft could carry a radar.

Stealth would change the nature of air combat. Stealth fighter pilots could acquire, identify, and select targets before their own aircraft were detected and fire missiles well outside visual range, and the target would have no idea that the chance of survival was down to about 10 percent. Col. Piccirillo compared the new fighter's use of stealth to the emergence of the U-boat in the 1914–1918 war: The objective is "to kill without being seen, disengage, and disappear," he said in late 1985. "The last thing you do is surface and use the deck gun. Close-in combat is something you try to avoid."

Fighter pilots are no more like the knights of romantic fiction than any soldier in history, the real peasant-whacking lance-jockeys of the Middle Ages included. The aerial duel, one pilot against another, is an unreliable ideal. The longer that a one-versus-one engagement lasts, the greater the chance that a third party will cruise into the arena and casually dispatch one of the contenders with a single shot. "There's a big luck factor in a dogfight," observed Piccirillo. "Ninety

Northrop's extraordinary YF-23 design changed barely at all between contract award in 1986 and its first flight in August 1990. Like Tacit Blue and the B-2, its stealth design combined a single continuous sharp edge that ran around the entire periphery of the aircraft, with curved and seamless upper and lower surfaces. *Northrop Grumman*

percent of people who get shot down, and come back, never saw who shot them down."

Lockheed and Northrop reckoned that they knew how to design a stealthy ATF that could fly as well as a conventional aircraft and would cost about the same, and the Air Force believed them. In November 1985, only two months after the RFP was issued, the Air Force released an amended requirement that placed more emphasis on stealth.

Meanwhile, things did not go well for the favorites. In 1985, the Electric Boat Division of General Dynamics hit the headlines in Congressional hearings that probed serious problems with the Navy's submarine contracts. General Dynamics went through some painful changes in senior management and was temporarily barred from Department of Defense work at a critical stage in the competition.

Some of McDonnell Douglas' senior executives were convinced that the SPO's emphasis on stealth was overrated and listened instead to F-15 pilots and Tactical Air Command officers, who were more skeptical about the new technology. McDonnell Douglas produced a disappointingly conservative design.

The stakes were raised in March 1986. Under pressure from Congress, the Air Force and Navy reached an agreement: The Navy's ATA would be designated as a replacement for the USAF's F-111, and an ATF variant would replace the Navy's F-14, making a joint requirement for well over 1,000 ATF's.

Another significant change followed the report of a commission headed by electronics-industry pioneer David Packard, who had been asked to look at reforms in Pentagon practices. The report came down heavily in favor of building prototypes of complex weapons; in May 1986, the Air Force announced that it would build and fly ATF prototypes. "That put us up another notch, because the Skunk Works is the best prototype shop in the world," a Lockheed executive recalls.

By now, it was clear that Air Force money alone would not support a winning Dem/Val effort. The winning contractors would have to contribute their own money to the project. Five of the competitors formed teams to share the load: Boeing and Lockheed with General Dynamics, and Northrop with McDonnell Douglas. The Lockheed/Boeing/General Dynamics team was formed in August 1986 on the basis that the team would collaborate on whichever design was selected.

In the final evaluation, Lockheed and Northrop led by a clear margin because they went further than their rivals in blending stealth with supersonic speed and agility. General Dynamics' jagged-edged delta placed third. Boeing came fourth with a V-tailed, shark-mouthed design, beating McDonnell Douglas into fifth place. Grumman and Rockwell trailed the pack. Overall, Piccirillo remarked later, "We had two excellent designs and three good designs, and two where the designers hadn't quite got the idea."

Lockheed's YF-22 and Northrop's YF-23 were declared the winners on October 31, 1986. Dem/Val was the largest fighter competition in history, lasting more than four years and costing almost $2 billion. Each team built two prototype aircraft. One of each pair of prototypes was fitted with the Pratt & Whitney YF119 and the other with the General Electric YF120. The teams built complete integrated avionics systems and flew them on test-bed aircraft, tested large-scale radar cross-section (RCS) models, and ran thousands of wind-tunnel hours.

Lockheed's design clearly resembled the F-117, with swept trapezoidal wing and tails—but, as the alert reader will note, this description does not resemble the YF-22. In July 1987, Lockheed junked its original design, which was judged overweight and uncompetitive. In a hectic three-month process, with help from its partners, Lockheed produced a largely different configuration with a clipped-delta wing.

Another change to both designs was confirmed during 1987. The only way to meet the very tough STOL requirement was to provide the ATF with thrust reversers that could be used in flight. In October 1984, McDonnell Douglas received a contract to modify an F-15 with vectoring and reversing engine nozzles; but by 1987, before the modified aircraft had flown, it was clear that the reversers would be heavy and complex—the F-15 nozzle included 5 miles of welds. To everyone's relief, the STOL requirement was scaled back and the reversers were eliminated.

The competition was largely conducted in

secret, to confound both the Russians and the media. (The Air Force's program manager during Dem/Val, Brigadier General James Fain, did not give a single on-the-record interview during his tenure.) The shapes of the rival designs were not revealed until summer 1990, when they were almost ready to fly. The program was running a few months behind schedule, partly as a result of Lockheed's redesign.

The YF-23's diamond-shaped wings and deeply recessed, nonvectoring nozzles indicated a concern that the RCS should be little larger from the rear and sides than from the front. This all-aspect stealth design was the fundamental difference between the two ATF contenders. The long forebody accommodated pilot, avionics, and weapons, and the engines were housed in widely separated nacelles. *Northrop Grumman*

The two winning designs were different from each other and from any previous fighter. Northrop's YF-23 was a truly alien-looking craft, with diamond-planform wings and huge V-tails. The F-22, by contrast, seemed surprisingly conventional, with four tail surfaces, vectored thrust, a broad, solid body, and a conventional wing.

Lockheed and Northrop had both run simulated combats and operational analysis and had come to different conclusions. Northrop believed that an ultrastealthy fighter could force a decision in the air battle before the hostile force could close to visual range. A few "leakers" would survive, but they would be outnumbered. Accordingly, the Northrop design stressed speed and stealth. The V-tails, deeply recessed exhausts, and jagged tail showed that it would have lower RCS from the side and rear than its competitor and a smaller infrared signature from below.

Lockheed, on the other hand, had deliberately sacrificed some stealth features to preserve controllability and maneuverability, improving its chances in a dogfight.

Northrop was first in the air, on August 27, 1990. The first YF-22, powered by the GE F120 engine, flew on September 29, piloted by Lockheed chief test pilot Dave Ferguson. The first F-22 went supersonic on October 25. The second F119-powered aircraft flew on October 30 with Tom Morgenfeld at the controls. A short, intense period of flight-testing followed—the two F-22s made 74 test flights in three months.

In early November, the first YF-22 sustained Mach 1.58 without afterburner. During December, it demonstrated its low-speed maneuverability, performing 360-degree rolls at a 60-degree alpha (angle of attack). Unarmed AIM-120 and AIM-9 missiles were fired from the side and center missile bays. The YF-22 flight-test program was completed on December 28.

The YF-23 did not go above 25-degree alpha in its tests; Northrop claimed that tests in both normal wind tunnels and spin tunnels tests showed that the aircraft could perform tailslides and had no alpha limits. Neither did it fire weapons.

As flight-testing continued, the competitors prepared their proposals for the engineering and manufacturing development (EMD) stage of the program. This included designs for two production versions—for the Air Force and Navy—and plans for development and production. Proposals were submitted on the last day of 1990.

An Air Force team evaluated the proposals (Lockheed's ran to 21,000 pages), but it was up to Air Force Secretary Donald Rice to make the final selection. The decision hinged not just on what the contractors promised, but on the customer's confidence in their ability to deliver. In early 1991, Northrop's partner, McDonnell Douglas, found itself embroiled in the collapse of the A-12 program;

U.S. AIR FORCE
ATF
YF-23
800

The Navy ATF (NATF) requirement called for a fighter that could carry high-speed anti-radiation missiles (HARM) and other weapons to clear a path for the stealthy but subsonic A-12 Avenger II bomber. This made it bigger than the Air Force design, so Lockheed offered the Navy an almost completely new swing-wing airframe, combining long endurance with carrier-compatible approach speeds. Engines, avionics, the forward fuselage, and stealth technology would be the same as the F-22. *Lockheed Martin*

Northrop was in trouble on the TSSAM missile. "The A-12 helped us," remarked the Lockheed official at the time, "because nobody's buying promises any more."

Lockheed's F-117, meanwhile, became the hero of the Gulf War. As a Lockheed executive put it, "They chose us because they believed us. When we thought that you couldn't do something, or more importantly that you couldn't do something for the money, we said so."

In several key areas, Lockheed had gone further in its demonstration program. The YF-22 had gone to high alpha and had fired missiles, and it had flown with a prototype advanced cockpit. The Lockheed team's avionics demonstration was more comprehensive.

Of the two proposed EMD designs, Lockheed's was more like what had flown in Dem/Val. The wing sweep was reduced and the span was increased to improve low-speed and maneuvering performance. The cockpit was moved forward and the inlets aft to improve the pilot's field of view over the nose and sides. The wingtip shape was changed to provide a better location for antennas. The vertical tails were shortened and the speedbrake was removed.

It was a double irony that Lockheed, which had won Dem/Val with a design different from the aircraft it built, scored points in the EMD selection for another design that would never see the light of day. Lockheed's Navy ATF (NATF) design "was an important factor," according to Air Force Secretary Donald Rice.

The NATF was due to start EMD in late 1993, with the first aircraft flying in January 1997. It was a fighter/strike aircraft. Lockheed proposed to blend the F-22 engines, avionics, and cockpit into a new swing-wing airframe. The Navy planned to buy 546 NATFs to replace F-14s. But after the A-12 was scrapped in January 1991, the Navy decided to divert NATF and A-12 money into the F/A-18E/F Super Hornet, followed by a new attack aircraft called AX.

The Lockheed, Boeing, and General Dynamics team was declared the winner in April 1991. The formal EMD contract was awarded in August, after a final review of the entire program.

The engine competition was just as intense. There was one fundamental difference between the two engines: General Electric's F120 was the first "variable-cycle" engine. For maximum power, supersonic acceleration, and supercruise, the engine was a pure turbojet, but at subsonic cruise the bypass ratio could be increased for greater efficiency.

General Electric emerged from ground-tests later than Pratt & Whitney. By then it was clear that the Dem/Val prototypes would weigh 60,000 pounds rather than the 50,000 pounds that had been the goal, so General Electric scaled up the YF120 to match. The General Electric-powered Dem/Val prototypes were faster, but the General Electric engine was not as mature as the Pratt & Whitney engine as Dem/Val ended. Pratt & Whitney argued that it could scale up the F119 to match General Electric's performance in a design that might be less complex and expensive and that presented fewer technical risks. Pratt & Whitney could also claim more experience with thrust vectoring, having been the contractor for the F-15 program. The F119 was accordingly selected for the F-22.

The contract called for the construction of 11 EMD aircraft, including two F-22B two-seaters. The first aircraft was due to fly in August 1995, and initial operating capability was set for 2001. The development and production plans had been revised under by the Pentagon's Major Aircraft

After the A-12 was scrapped, the Navy dropped the NATF as well, planning instead to build a new dual-role aircraft known as A/FX. The Lockheed/Boeing/GD team proposed a swing-wing aircraft based on F-22 technology, but with two seats and less powerful PW7000 engines derived from the F119. The Air Force was interested, but the Clinton administration ordered cuts in the Pentagon budget that required the Navy to choose between the Super Hornet and the A/FX, and the latter was canceled in 1993. *Lockheed Martin*

Review, published in April 1990. Worried about its ability to afford all the major aircraft programs that would reach production in the 1990s, the Pentagon delayed ATF production by two years and cut the annual production rate of the Air Force version from 72 to 48 aircraft. A year later, Air Force Secretary Rice also announced that total production would be cut from 750 to 648 aircraft, in an early post-Soviet budget cut.

Lockheed announced before the EMD contract was awarded that, if it won, it would locate the new program's headquarters in Marietta, north of Atlanta, Georgia, where the forward fuselage also would be built. Marietta offered lower costs and a larger labor force than California, and Lockheed operated a massive production facility there. General Dynamics would build the midbody section in Fort Worth, Texas, and Boeing would manufacture

The McDonnell Douglas ATF design resembled Lockheed's concept in its wing and tail configuration—the St. Louis company had hired some Lockheed people to bolster its stealth skills—but had a wedge-shaped, divided chin inlet. Its stealth shaping features were a unique combination of faceted and curved surfaces. It was one of the heavier ATF designs, and its performance was not promising. McDonnell Douglas was shocked to be placed fifth in the evaluation, behind Boeing. *McDonnell Douglas*

the wings and tail, develop the training system, and perform the critical task of avionics integration.

But as this book is written, in 1997, the F-22 is eight years from entering service. When that happens, the newborn son who the author left to attend that Anaheim conference in 1982 should be in his last year of college. Anything that takes that long should be worth the effort—and a detailed look at the F-22's technology shows why its supporters think that is the case.

Chapter Two

Stealth and Supercruise

The F-22's basic shape, devised in three hectic months in 1987, faced a fundamental challenge: to reconcile the demands of stealth, supersonic cruise, and agility.

Stealth influences the shape and angle of all external surfaces. It requires that all weapons and fuel be carried internally, demanding an airframe of much greater volume than a nonstealth design of the same performance.

Supersonic cruise requires low supersonic drag, which in turn implies slenderness and thin wing and tail sections—neither of which are inherently compatible with large volume.

Agility is achieved through effective controls and a large wingspan and area. This is hard to reconcile either with the need for a small, thin wing for supercruise, or with the fact that the best tail for a stealth aircraft is no tail at all.

One of these challenges in developing the F-22 was the design of a new generation of compact, flush antennas that would not compromise the fighter's stealth characteristics. Here, a representative upper fuselage segment with a buried antenna is tested in a Lockheed Martin anechoic chamber. *Lockheed Martin*

The F-22 is about the same size as an F-15, but is heavier, tipping the scales at some 62,000 pounds in clean condition. In general layout, the F-22 is a moderately swept (42-degree) delta. The delta wing combines low thickness/chord ratio for supersonic drag with enough area to meet maneuverability requirements. It offers a useful amount of space for fuel, and its span is still short enough to fit in standard NATO aircraft shelters. The wing and body are highly blended to add volume.

The Raptor is a fly-by-wire aircraft, with the hydraulic actuators that move its aerodynamic surfaces controlled by electronic signals generated by computer. Fly-by-wire, however, is not magic, and it can only perform within the limits set by the authority of the control system and the airplane's physical responses. The designers' goals were ambitious and would be achieved by a combination of the software in the fly-by-wire computers and effective aerodynamic control.

The F-22 was designed to be agile, reflecting a slightly more complex concept of maneuverability than the philosophy that lay behind earlier fighters. It would match or exceed them in basic

The F-22's densely packed interior is apparent in this "ghosted" view. This image was generated using the Dassault Systemes CATIA computer-aided design system, a French-developed program that has become an industry standard. Different colors identify different systems: purple for weapons, green for propulsion, orange for hydraulics, and so on. *Lockheed Martin*

parameters such as sustained turning or rolling rates, but it would also be able to move more quickly from one maneuver state to another—for instance, from a high-G turn to a straight-line acceleration—and would be able to reach high alpha (angle of attack) while remaining under full control. Another goal was to avoid stability and control deficiencies that require the flight control system to set limits on alpha, or to restrict other maneuvers (such as roll) at high alpha.

The designers aimed for "carefree abandon" handling, allowing the pilot to exploit a large alpha/airspeed envelope without overstressing the aircraft or causing it to depart from controlled flight. The F-22 is designed to be immune to deep stalls—stalls from which the aircraft cannot recover with normal control inputs—and to recover from high-alpha, post-stall conditions with both engines flamed out. Lockheed chose a four-tail configuration because it provides stability and

Almost 17,000 hours of wind tunnel testing were performed during the EMD program alone, involving 23 models in 15 facilities in the United States and Germany, and following a roughly equal number of hours on the Dem/Val design. Remarkably, no significant changes have been made as a result of tunnel tests. This all-metal model is performing high-speed tunnel tests. *Lockheed Martin*

Low-speed, high-alpha characteristics are always the hardest to predict, because of the mix of smooth, turbulent, and vortex flow. This model was specially built for high-alpha work and is supported on a ball-jointed pylon so that it can be moved to different pitch and yaw angles. Smoke generators and sheets of laser light can be used to look precisely at the behavior of the flow. *Lockheed Martin*

linear control response in pitch, roll, and yaw over a wide speed and alpha range.

The F-22 is the first fighter to be designed from the outset to use vectored thrust for control. Thrust vectoring is not used to expand the flight envelope, because the Air Force insists that the F-22 must be able to recover without it. Rather, thrust vectoring gets the F-22 from one maneuver state to another more quickly. The same benefits could have been achieved with conventional controls, but it would have meant increasing the size of the tails by 30 percent, adding 400 pounds to the empty weight and increasing drag.

The flight control system (FCS) operates the horizontal tails, the rudders, the vectoring nozzles, the wing surfaces (flaperons, ailerons, and leading-edge flaps), and even the nosewheel steering. There are no speedbrakes: for in-flight deceleration, the flaperons go down, the ailerons deflect up, and the rudders move outward. On the ground, the entire trailing edge deflects up to spoil the wing lift. Flight and propulsion control are fully integrated.

The first F-22A flew with a set of FCS laws that will address the full flight envelope and all configurations, according to Lockheed Martin. This is a contrast with some other programs, such as Eurofighter, where early flights were made with provisional software that set tight limits on alpha and maneuvering rates. Although testing will be incremental, YF-22 and wind-tunnel experience suggests that no major changes will be necessary.

The F-22's stealth design evolved from that of the F-117, with a preponderance of flat, canted surfaces and a sharp chine line from the nose to the wingtips. Most surfaces are swept and sloped to deflect radar energy away from any radar that illuminates the aircraft. Residual reflections from edges and apertures are concentrated on a few alignments—doors and panel junctions are serrated, their edges parallel with wing and tail edges. A basic difference between the F-117 and the F-22 is that radar-absorbent material (RAM) is not applied to the entire aircraft, but used selectively on edges, cavities, and surface discontinuities. Lockheed Martin builds all the edges of the aircraft, which probably consist of wide-band radar-absorbent structures with honeycomb or foam-type internal structures, doped with lossy

A propulsion-systems model on test in the low-speed wind tunnel at Lockheed's Marietta, Georgia, plant. Smoke is being generated around the inlets to ensure that the air will flow smoothly to the engines when the airplane is at low speed and the engines are at full power. *Lockheed Martin*

ingredients, which convert radar energy to heat. Heat-resistant ceramic-matrix RAM is likely to be used on the exhaust nozzles. The radome is a "bandpass" type that reflects signals at all frequencies except the precise wavelengths used by the F-22 radar.

One of the principal challenges in the development of the F-22 is the development of RAM which can withstand the temperatures and pressures encountered in supersonic flight, is durable, and requires little maintenance. A key requirement for the F-22 is that it will be maintained outdoors, like any other fighter. Despite reported problems with the B-2, Lockheed Martin is confident that this requirement will be met.

Radar cross-section problems were discovered during early full-scale model tests. The problem was traced to the difficulty of maintaining

Final assembly of the first F-22A begins with the forward fuselage. This section, which forms the fuselage just behind the cockpit, is mostly aluminum because it does not experience peak structural or thermal loads. Four long structural members—composite side beams and aluminum longerons—connect the front fuselage sections. *Lockheed Martin*

tolerances in a large number of apertures and serrations. As a result, many access panels and drain holes were eliminated or combined, and some serrated edges were modified with fewer, larger teeth. Tests of a modified full-scale RCS model "on the pole" at Lockheed Martin's state-of-the-art RCS range at Helendale, California, have indicated that the problem is solved.

The F-22 bucks the trend toward the use of composite materials. Only 35 percent of the bare airframe is composite, although the designers had expected to use more than 40 percent. The main reason was that the specialized composites needed to withstand high temperatures caused by skin friction at high speeds were expensive and hard to work with.

The designers nevertheless reached their weight goal—25 percent lighter than an all-aluminum airframe—by using a lot of titanium, which makes up 41 percent of the airframe weight. Composites are found mainly in the skins, and in the wings and tails where their stiffness is valuable.

The heart of the structure is the massive and complex midbody section, built by Lockheed Martin Tactical Aircraft Systems in Fort Worth. It incorporates the four weapon bays, the main landing gears, and the complex inlet ducts. Made of carbonfibre/epoxy, the ducts curve sharply upward and inward from the inlets to mask the engine faces from radar, changing section smoothly from rhomboidal to circular. Their inner contours must be smooth and accurate to maintain their stealth characteristics. Attached to the midbody are the forebody, accommodating the cockpit and avionics, which is built by Lockheed Martin in Marietta, and the wings, aft fuselage, engine bay, and tailbooms, built by Boeing.

The midbody and rear fuselage include some

Forged and machined titanium "spectacle" frames knit together the aft-fuselage structure, built by Boeing in Seattle, which accommodates the engines and supports the horizontal tails on twin booms. Titanium is used in this area because it can withstand the very high temperatures generated by their engines and exhausts. *Boeing*

unusual structural features. The inlet lip and the fittings that support the wing and rudder are hot isostatic process (HIP) castings, made from titanium alloy powder formed under very high pressure. HIP was originally developed for disks in engines, but is used to form highly loaded, rigid, complex-shaped components—the only other way to build them is from many small parts. The tailbooms are titanium, welded by an electron beam in a vacuum chamber. The aft fuselage is 67 percent titanium because of high temperatures.

Carbonfibre/bismaleimide (BMI) composite is the primary material in the wings. The BMI resin replaced the thermoplastic-matrix composite

The aft fuselage is clamped into a tool and tipped on its side, ready to be loaded into a data-driven, laser-guided drilling machine which will drill, countersink, and deburr 2,000 holes into the structure. The holes will be used to attach the composite upper skin and the lower engine-bay doors. *Boeing*

used in the YF-22 because it was stronger and less expensive, and because tougher, more damage-tolerant BMI resins had become available during Dem/Val. Thermoplastics tolerate higher temperatures than BMI, so the change to BMI in the EMD aircraft meant a reduction in maximum Mach number, from 2.0 to 1.8.

The Sierracin canopy is unique in that it is a one-piece bubble with no arch or bow. Most canopy specifications require near-perfect optics only in the forward field of view, but the F-22 will have a helmet-mounted sight and therefore needs "zone 1 quality" throughout.

Birdstrike protection remains an issue. The F-22 canopy is not as inherently tough as the multilayer F-16 canopy. Although the F-22 canopy can withstand a 450-knot birdstrike, the impact initiates a wave through the canopy which, at its lowest

The massive midbody section is assembled by Lockheed Martin Tactical Aircraft Systems at Fort Worth. Here, the front and rear sections of the midbody are being joined, viewed from the rear. The forged and machined titanium bulkheads carry most of the flight loads, and the largest of them weighs more than 3 tons before machining. Note that many fuel lines and accessories are already installed. *Lockheed Martin*

point, strikes the heads-up display (HUD) combiner, sending fragments into the pilot's face. HUD supplier GEC Avionics and Lockheed Martin are working on designs for a collapsible combiner.

The F-22 was designed and built differently from most earlier aircraft. First, it was entirely designed on computers, eliminating drawing boards and mockups. One of the major advantages of this approach is that every measurement on the aircraft is contained in a computer database. Every part of the aircraft is designed according to the database, and the gauges and other tools on the production line that measure the parts built are also linked to the database. The result is that even large, complex subassemblies conform precisely to the database and fit near-perfectly. Witnesses report an

A mutant banana? No, an Alliant TechSystems graphite-epoxy composite pivot shaft that supports the horizontal tail. Composites provide the high stiffness-to-weight ratio that makes this heavily loaded component practical. Its complex shape—circular at the trunnion and flattened where it attaches to the tail structure—is created by winding graphite fiber, up to 490 plies thick, around a core mold or mandrel while the part is sprayed with matrix material. *Lockheed Martin*

audible "click" as fuselage parts go together.

The Pratt & Whitney F119-PW-100 engine was the only part of the F-22 developed ahead of the airframe, a measure of the challenge that it presented. It is the world's most powerful fighter engine. Although its thrust is officially quoted as "in the 35,000-pound class," the actual figure may be more than 39,000 pounds with full augmentor, implying an intermediate (nonaugmented) rating of 25,000 to 27,000 pounds.

But static thrust is only part of the story. Pratt & Whitney has said that at supersonic speed, on intermediate thrust, the F119 generates twice as much power as the F100-PW-200. This was the critical requirement from the start of the ATF program. Today's fighter engines are efficient at subsonic speeds, but can propel aircraft supersonically only by burning extra fuel in the augmentor. This was the only way to provide the thrust required for supersonic acceleration and combat in an engine that was light enough to allow a fighter to be agile; but in full burner, the engine consumes between two and three times as much fuel for every pound of thrust as it does without

Here, the skins are mated to the Boeing-made wings. The multispar wings incorporate sine-wave spars—in which the web is an undulating curve—produced by a resin-transfer molding (RTM) process developed by Boeing and Dow/United Technologies. In the RTM process, dry carbonfibre fabric is laid up in a mold, and BMI resin is injected at high pressure. One in four of the spars is still made from titanium—visible by its green-primer finish—a change made after live-fire damage-tolerance tests. *Boeing*

the augmentor. This limits the use of maximum thrust to minutes.

The only way to sustain supersonic speed was to do so on intermediate thrust, without the augmentor. This was not just a matter of a more powerful engine. A jet engine's inlet decelerates the high-speed airflow entering the intake and compresses it. When the air is compressed it gets hot, and the temperatures inside the engine increase as well. Around Mach 1, in today's engines, the turbine reaches its maximum temperature, and at higher speeds the engine is automatically throttled back. The only source of more power is the augmentor.

The ATF engine would have to run at higher temperatures and it would have a lower bypass

Boeing workers complete the rear-fuselage assembly before it is shipped to Marietta. All electrical, fuel, and hydraulic lines are already installed and simply need to be connected when the sections are mated. Invented in wartime Germany and rediscovered and refined in the 1970s by Airbus Industrie, this technique makes final assembly simpler and more efficient. *Boeing*

ratio. That is, almost all the air would flow through the core—the compressor, combustor, and turbine. Even though the engine might not be much bigger than the F-15 and F-16 engines, it would have a much larger and hotter core. This would present a challenge in terms of weight, because the core is the heaviest part of an engine.

Of the many new features that the F119 engine embodies to meet those requirements, the most basic is the use of computer models that predict the effects of airflow to design compressor and turbine blades. Without these, designers could not predict the aerodynamics of a blade at its ends, where it meets the disk and case of the engine, so they minimized these "end-wall" effects by using long, thin, straight blades. Now, engineers can produce thicker, shorter, twisted blades that can do more work.

The midbody of the first F-22 is prepared for delivery. The entire midfuselage structure is a fuel tank, apart from the inlet ducts and the bays that accommodate the landing gear, weapons, refueling slipway, gun, and part of the environmental control system. This design feature allows the clean F-22 to carry at least as much fuel as an F-15 with three 600-gallon tanks. Note the serrated skin joints, which reduce the need for tapes and seals to maintain stealth. *Lockheed Martin*

This leads to a virtuous spiral of improvements. There are fewer blades on each stage. The blades are thicker, so the disk-to-blade junction is wider and stronger. The disk can spin faster without the risk of failure and can do more work. This means that each row of blades, or stage, can generate a greater rise in pressure, and the number of stages in the engine can be reduced. The engine is shorter, lighter, and stiffer, and has fewer parts.

The turbine, too, has fewer stages and blades. This is important because the gas in the turbine is hot enough to melt any metal, so each blade has a labyrinth of tiny air ducts that carry away heat from the metal and shed a film of cooler air over the surface of the blade. If there are fewer turbine blades, the cooling air supply can be more concentrated, and the thicker blades can be designed with more efficient internal cooling.

Unlike earlier large engines, the F119 does not have separate blades and compressor disks. All the stages except the first-stage fan are "blisks" with blades and disks in one piece. The first stage

The structurally complete forward fuselage is loaded into an assembly tool where avionics racks, wiring, and the cockpit interior will be installed. Radar absorbent structure (RAS) will be attached to the flat side surfaces to complete the chined shape. *Lockheed Martin*

of the fan has massive, wide-chord blades, made of hollow titanium, which are more efficient, lighter, and more robust than the F100's blades. The fan blades are made separately and joined to the disk by linear friction welding, a technique in which the blade is rubbed so hard against the disk that it bonds to it.

Progress has not been completely smooth. Soon after the engine made its first runs in the end of 1992, Pratt & Whitney discovered a vibration problem in the transonic-flow, counter-rotating turbine. A new turbine was designed and built in 18 months and made its first run in May 1995. The engine had undergone 5,000 hours of testing before first flight and is not expected to require anymore changes before the F-22 enters service.

The F-22 incorporates innovations in unglamorous, often overlooked but very necessary subsystems. It is the first Air Force fighter in many years to have a specially developed life support system (produced by Boeing). It is designed to reduce the risk of G-induced loss of consciousness (G-LOC). Unlike the familiar symptoms of "blackout" at high G, G-LOC causes sudden and complete loss of consciousness, followed by a recovery that can take minutes. It emerged as a problem with the arrival of the F-16 and other high-performance aircraft, because it is associated with rapid increases in G loadings.

The F-22 life support system reduces the risk of G-LOC in several ways. It includes a new anti-G garment that covers more area than earlier G-suits, so that it can exert pressure on more of the body's blood supply and raise blood pressure in the head more effectively. The oxygen system can pump oxygen into the pilot's lungs under pressure, increasing the oxygen content in the blood. This requires a special oxygen mask and a counter-pressure vest that prevents the pilot's chest from emulating a pigeon. The G-suit and positive-pressure system are controlled by a breathing regulator and anti-G garment (BRAGG) "smart valve" that reacts to the rate of G onset. The system also includes the HGU-86P helmet, developed by Helmets Integrated Systems of the United Kingdom, which is lighter than previous helmets and is designed to fit better.

Air Force tests have shown that positive-pressure breathing, the smart valve, and improved anti-G suit not only reduce the risk of G-LOC but improve G tolerance and allow the pilot to sustain G with less physical strain and fatigue, an important factor in sustaining high sortie rates.

Positive-pressure breathing also provides altitude protection. Air Force fighters are normally limited to a maximum altitude of 50,000 feet

The forward and midfuselage sections are mated, and the aft fuselage is lowered into position. In the past, the joining of major body sections was a tricky operation, requiring a great deal of adjustment and the use of assembly shims. Computer-aided design and manufacture permits a dramatic increase in precision, so that parts almost snap together. *Lockheed Martin*

because, if power and cockpit pressure are lost, the pilot will lose consciousness before the aircraft descends into thicker air. The F-22 life-support ensemble has been chamber-tested to 66,000 feet, and its emergency oxygen system will function long enough for the pilot to reach lower altitudes.

The life-support system includes an air-cooling garment underneath the G-suit and counter-pressure vest, and optional suits that protect the pilot from chemical and biological agents and cold water immersion.

The pilot is not the only system that must be kept cool. Conventional air-cycle environmental control systems (ECS), which cool the cockpit and the avionics, have two drawbacks for the F-22: they require cool incoming air, which may be hard

Aft fuselage mating continues. Each station on the F-22 final assembly line includes docks and elevated work platforms that fit around the aircraft, so that workers can reach all parts of the aircraft easily without perching on ladders. *Lockheed Martin*

The complete F-22 fuselage is ready to be lowered into position for wing mating. Note the cavernous side weapon bays—large enough for the standard AIM-9L Sidewinder—and the stealth-influenced rear slope of the radar bulkhead. *Lockheed Martin*

to come by at Mach 1.5, and their exhausts are infrared hot spots. The F-22 ECS uses air-cycle cooling only for flight-critical avionics. This system also feeds the onboard oxygen-generating system (OBOGS), which provides the pilot with oxygen. Mission avionics, which impose higher cooling loads, are liquid-cooled, and the coolant flows through coolant-to-fuel heat exchangers. The heat is either dumped overboard as fuel is burned or removed through an air-cooled heat exchanger on top of the fuselage.

The ECS and other systems are important in achieving another F-22 goal: cutting both the cost of maintenance and the number of people and tonnage of equipment needed to support the fighter in the field. For example, the use of OBOGS simply eliminates liquid oxygen, a major element of the logistics chain.

Much of the improvement in maintainability comes from a change in design philosophy. The classic process of aircraft design was serial. The aircraft was designed by one group of engineers. As they completed the design of each part, they passed it to manufacturing engineers, who would work out how to make and assemble the parts.

The first F-22 gets its wings. This is the final major mate operation before moving surfaces, the vertical tails, doors, landing gear, and RAS edges are installed. This view shows clearly that the F-22 configuration is similar to a tailless delta, with the addition of twin booms to carry the horizontal tail surfaces. *Lockheed Martin*

The F-22 canopy, made by Sierracin, comprises two 0.4-inch (9.5 millimeter) sheets of polycarbonate, sandwiched between two sheets of optical glass, fusion-bonded in an autoclave, and drape-formed over a canopy blank at 750 degrees Fahrenheit. A metallic coating of indium-tin oxide is added to the canopy to reflect radar waves, giving it its gold tint. *Lockheed Martin*

When the aircraft reached the user, the maintainers would learn how to repair and service it.

The result was usually something that worked, but which cost more than it needed to. The original design would prove expensive or difficult to make and would have to be changed. Maintenance would be difficult because of design features that would be obvious to a maintainer on the ramp but had been overlooked in the original design. Production difficulties would mean that parts would have to be removed from the line and reworked or modified to fit.

The F-22 was designed by integrated product teams (IPTs), each responsible for one part or aspect of the system. The IPT included design engineers, manufacturing specialists, and maintainers, and no part could be released until all their requirements were met. At the same time, the IPT also generated all the documentation that would follow each part through production and service, describing the process for making and repairing it.

Other changes include built-in test equipment that replaces off-board test sets, and more items designed to be replaced on the flight-line rather than repaired in an intermediate-level shop on the base. A 24-aircraft unit of F-22s requires only 8 C-141B loads of equipment for a 30-day deployment, versus 18 for the same number of F-15s and half as many people. An essential part of the maintainer's equipment will be the Portable Maintenance Aid, a rugged notebook-type computer that replaces paper manuals.

Armament was one of the controversial issues throughout the ATF program. A stealth aircraft needs internal weapons, but the standard Air Force weapon, the GM-Hughes AIM-120 Advanced Medium-Range Air-to-Air Missiles (AMRAAM), was not designed for internal carriage and its wing and tail surfaces make it a bulky package. One answer was to develop a version of AMRAAM with folding wings, but it might not work well when carried externally, and the Air Force emphatically did not want a split inventory of ATF and non-ATF missiles.

The solution is a new version of AMRAAM, the AIM-120C, with clipped wings and tails. Its performance is virtually identical to earlier AMRAAMs, and it will be the standard version for all Air Force fighters, but it takes up considerably

The Pratt & Whitney F119-PW-100 engine is similar in dimensions to current fighter engines, with about the same airflow as the F100 (about 275 pounds per second), but its bypass ratio—the proportion of the air that bypasses the compressor and feeds the augmentor—is 0.2:1 or less, versus 0.7:1 for the F100, so the compressor, combustor, and turbine are at least 50 percent bigger. The geodesic pattern on the case is produced by chemical milling. *Pratt & Whitney*

The F119 has a three-stage fan, a six-stage compressor, and single-stage low- and high-pressure turbines. (The F100 has 10 compressor stages to achieve the same 25:1 pressure ratio.) The low-pressure and high-pressure shafts spin in opposite directions: this eliminates a stator, reducing the engine's length and cutting the requirement for cooling air. Each of the high-pressure turbine blades, small enough to fit in the palm of your hand, extracts 1,000 horsepower from the gas stream—equivalent to two race-tuned Corvette engines. *Pratt & Whitney*

An F119 runs at full burner during tests. New features of this hot-running engine include "dual heat treat" turbine disks: the center of the disk, where stress loads are concentrated, has a fine-grain crystalline structure for maximum strength, while the rim has a coarser grain for better damage tolerance. The Floatwall combustor, with a "shingled" design to reduce stress, is expected to last 10 times longer than the state-of-the-art rolled-ring design. *Pratt & Whitney*

Early in the F119's development, Pratt & Whitney engineers donned chemical and biological warfare ensembles and joined operational Air Force F-15 maintainers on the flight-line. As a result, the designers selected a small set of wrenches, ratchets, and sockets and built the engine so that all exterior maintenance could be carried out with those tools and restricted themselves to a few types of clips and fasteners. Virtually all the engine's plumbing is accessible without removing the engine itself, and all lines are color-coded. *Pratt & Whitney*

less space. Three missiles are carried in each of the F-22's ventral bays, which are covered with bi-fold thermoplastic-composite doors. The AIM-120s will be propelled from the weapon bays by pneumatic/hydraulic AMRAAM Vertical Ejector Launcher units.

The side bays will each hold one GM-Hughes AIM-9X Sidewinder, carried on the AIM-9 Trapeze Launcher, a mechanically extending rail incorporating an exhaust plume deflector. The AIM-9X is a radically modified version of the veteran missile, with thrust-vector control and a seeker capable of locking onto targets 90 degrees off the fighter's boresight. The Trapeze launcher will extend automatically as the F-22 nears the point of achieving launch parameters on the target, allowing the infrared seeker to lock on before launch.

At one point in the evolution of the ATF, as the requirements for the production aircraft were being completed, the SPO raised the question of eliminating the gun to save weight. A howl of protest arose from the fighter community: the F-22 has a gun. Although the number of gun kills in air combat has declined precipitously since new-generation missiles such as the AIM-9L Sidewinder entered service in the late 1970s, pilots argue that the gun is the only effective weapon against an adversary who has forced the

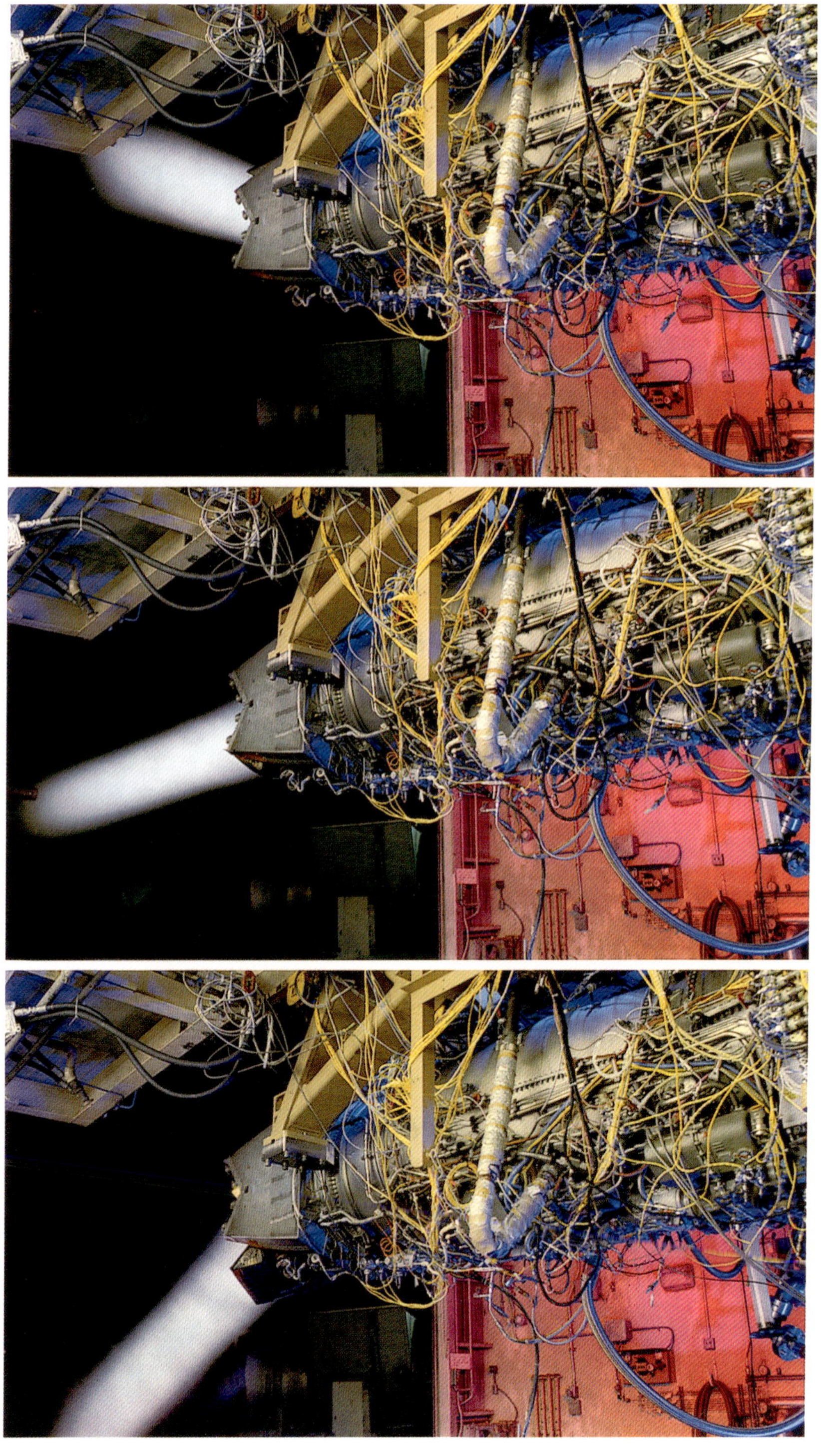

The vectoring nozzles can divert the full augmented thrust 20 degrees upward or downward in a second. Two-dimensional nozzles are necessary for stealth: The edges of a 2D nozzle can be aligned with the other edges of the aircraft, and its shape flattens the exhaust plume and promotes mixing with the ambient air. The nozzles are largely made of burn-resistant Alloy C titanium and incorporate a sophisticated internal cooling system. *Pratt & Whitney*

fight inside the minimum range envelope of the missile. Also, the gun has been made more effective by improvements in ammunition, gun-aiming software, and HUD symbology.

The Air Force asked Lockheed in 1994 to develop an air-to-surface capability for the F-22, and the lower weapon bays were modified to accommodate the 1,000-pound Boeing GBU-32 Joint Direct Attack Munition (JDAM). The F-22 can carry two JDAMs, two AMRAAMs, and two AIM-9s. A synthetic aperture radar (SAR) mode is being added to the F-22's radar for air-to-surface operations.

Two 1,000-pound bombs may not seem like much, but with SAR and JDAM the F-22 will be able to deliver weapons within 30 feet of a target in almost any weather. The Air Force plans to develop a

The F-22 will have a paperless logistic system. The Portable Maintenance Aid (PMA) replaces a cartful of paper manuals and plugs directly into the F-22's diagnostic systems through ports in the cockpit or main landing gear bay. The PMA and other definitive logistics tools will be used throughout the test program, so that they will be mature when the F-22 enters service. *Lockheed Martin*

low-cost programmable seeker for JDAM, which will make it as accurate as a laser-guided bomb but autonomous and unaffected by most weather.

With the advent of the F-22 and Joint Strike Fighter, any new-generation weapons under study are being designed for internal carriage. They include a follow-on for the Raytheon AGM-88 High-Speed Anti-Radiation Missile—a ramjet-powered, Mach 6 weapon with a 100-nautical-mile range and a dual-mode seeker. Boeing is working on the Miniaturized Munition Technology Demonstration, a small hard-target guided weapon weighing a mere 250 pounds but able to punch through 6 feet of concrete. The F-22 could carry eight such weapons.

When stealth is not critical, the F-22 can carry up to 5,000 pounds of external stores on each of four underwing pylons. For ferry flights, each of these can accommodate a 600-gallon fuel tank and a pair of AMRAAMs, reducing the need for tanker and cargo support, or the F-22 can carry two tanks and two dual AMRAAM launchers for defensive missions after the enemy air force is overwhelmed.

Flight tests should confirm that the F-22's conservative looks belie its performance. Details are classified, but a chart published in 1991 shows that the F-22 is faster on intermediate power than an F-15C on full burner, when both aircraft have eight AAMs on board. (The speeds are probably around Mach 1.6–1.7.) "We expect that this will be one of the things that surprises the Air Force," says Lockheed Martin test pilot Paul Metz. "If you don't know what you're doing, you'll be supersonic." Unlike most fighters, too, the F-22 achieves its highest rate of climb at supersonic speed.

The F-22 is nearly as fast with afterburner as without. The augmentors will be used mainly for acceleration and supersonic maneuvering. Metz believes that the "afterburner will generally not be required," and that when it is used it will be in bursts of seconds and tens of seconds, at the outside.

The maximum speed—between Mach 1.8 and

A Lockheed Martin engineer uses an optical-electronic "step-gap measurement tool" to check the fit of an access panel of the F-22. Panels and doors must fit within certain tolerances to maintain the fighter's stealth characteristics, and physical measurement of the entire panel is extremely time-consuming. A similar system will be deployed with operational fighters for use in the field. *Lockheed Martin*

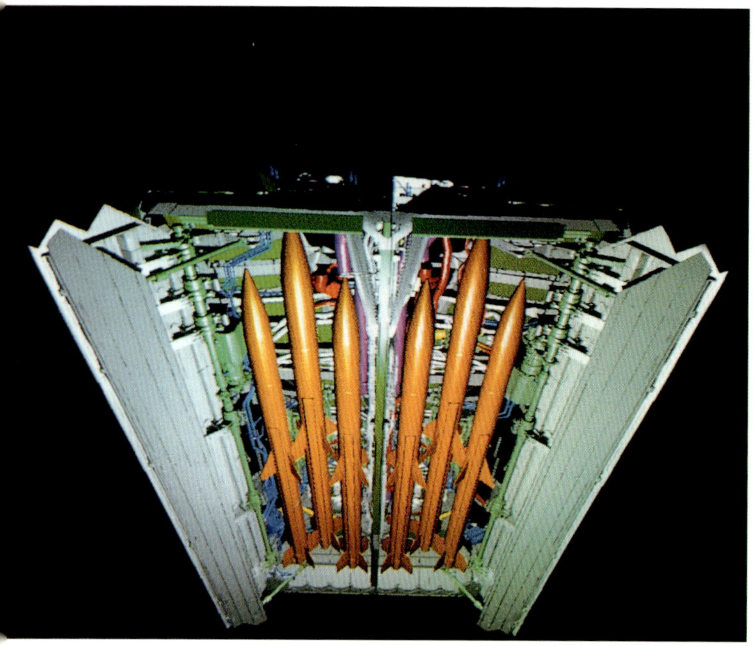

Three AIM-120C missiles fit in each of the F-22's two ventral weapons bays, covered by bi-fold doors. The central missile in each cluster is staggered above and ahead of the other two, so that the wings and tails overlap. The pneumatic-hydraulic launcher punches the missile out of the bay with a force of 40 g. *Lockheed Martin*

Mach 2.0—is lower than the nominal maximum speed of the F-15. But the F-15 can attain its top speed only with a minimal weapons load and no external fuel, and most pilots never see 2.5 on the Machmeter. The F-22 has plenty of thrust for Mach 2.5, but providing that performance would have demanded variable inlets and higher-temperature materials throughout the aircraft.

The principal breakthrough in terms of straight-line performance is supercruise. The Air Force has stated that "about 30 minutes in a one-hour mission" can be flown at supersonic speed, three to six times the supersonic endurance of any fighter using augmentors. On a typical mission, the F-22 can sustain supersonic speed for most of the time that it is over hostile territory. Thirty minutes at Mach 1.5 is equivalent to 250 miles each way. Supersonic endurance varies with speed: a supercruising F-22 may vary its speed between Mach 1.1–1.2 and Mach 1.5–plus according to the tactical situation.

Supercruise goes along with high altitude. Cruising higher and faster than other aircraft, the F-22 will give its pilot control of the engagement, and it is more likely to surprise its targets from the rear.

The F-22's reduced head-on RCS is claimed to guarantee a first-look, first-shot advantage against any contemporary fighter. Most fighters today are designed with reduced RCS from the head-on aspect. However, where the F-22 differs from any other air-combat fighter is in its all-round RCS, which is described as "in the birds and bees class"—that is, in the same order as the F-117 and B-2, and enough to reduce radar-detection range by a factor of 10 compared with conventional aircraft. The result is that the F-22 may be detected only transiently by the best air defense network, giving its pilots the advantage of surprise.

Stealth, supercruise, and high altitude work together to reduce the F-22's susceptibility to SAMs. The aim is to reduce detection range to the point where the SAM cannot complete an engagement successfully. By the time a SAM radar detects an F-22, it is too late for a head-on intercept: The fighter will be overhead and the range will be increasing before a missile can reach the fighter's altitude. In a tail-chase, the missile is low on energy and can be evaded without too much difficulty.

Some F-22 critics have talked about "radar stealth," implying that even if the F-22 evades tracking by radar it will still be detected by infrared (IR) sensors. But the F-22 was designed according

An F-22 releases two 1,000-pound Boeing GBU-32 JDAM guided bombs. JDAM is a conventional bomb fitted with a guidance system that combines an inertial sensor with a Global Positioning System receiver. Immediately before the bomb is released, it is programmed with the trajectory that the aircraft's weapon computer predicts that it will follow. The guidance system simply keeps it on that track, eliminating errors caused by winds and other factors. *Lockheed Martin*

to a philosophy of "balanced observables," which mandated that the F-22's IR signature be reduced so that IR and radar sensors would have a similar detection range. The most prominent source of IR radiation from an aircraft is its exhaust plume. On the F-22, plume radiation is reduced by minimal afterburner use, the mixing of the core and bypass flows, and by the 2-D nozzles, which create a flattened exhaust plume with a wider perimeter than a circular plume. This causes the plume to dissipate more quickly.

Much of the remaining IR signature comprises reflected solar IR radiation and emissions caused by skin friction heat. IR-absorbent paint

The F-22's gun is probably the oldest part of the airplane, dating back to a weapon first fired in the 1940s. The General Dynamics M61A2 20mm six-barrel Gatling cannon is a lighter version of the M61 with slimmer barrels and a redesigned breech. It is mounted above the right wing root, and the muzzle opens onto a shallow trench in the fuselage, covered by a side-hinged door. The F-22 carries 480 rounds of ammunition in a linear feed system, aft of the weapon bays. *General Dynamics*

reduces solar reflection; it is analogous to normal paint except that it absorbs in the IR band. Friction heat cannot be absorbed, but coatings have been developed that change the emissivity of a surface; that is, they make it less efficient at emitting IR. They may also be able to concentrate the wavelength of remaining IR energy into wavebands that attenuate most rapidly in the atmosphere and are therefore harder to detect.

The future of visual-range air combat has been a controversial issue throughout the F-22 program. As pilots put it, "If you have a rifle with a telescopic sight, and you're trying to kill a midget armed with a knife who's standing in a phone booth, the last thing to do is get into the phone booth with the midget."

The argument is sharpened by the spectacular low-airspeed, high-alpha maneuvers demonstrated by Russian fighters, culminating in the vectored-thrust Sukhoi Su-37. Some critics suggest that the F-22 represents a move in the wrong direction and that it is the Russian design that points to the future of air combat. Other observers point out that dumping speed and yanking the nose around make it possible to take a fast shot at an adversary, but at the same time, it makes you a big, fat, slow target for anyone else in the sky. You have killed your energy, you are not going fast, and for the next few seconds the only direction you are going is forward and down. "It's a neat trick, but it's of no use in air combat," is the assessment of one U.S. fighter commander.

Whichever way the argument turns out, the F-22 should be able to match the agility of any other aircraft in service or under development. Its flight envelope is very large: alphas as high as 60 degrees were demonstrated in the YF-22 program, and some roll maneuverability was retained at that extreme pitch angle. At alphas of 15 degrees and above, the F-22 rolls at least twice as fast as the F-15, and the gap widens until the F-15 hits 30 degrees alpha and can no longer roll at all.

The F-22 is able to get around its envelope quickly. Maximum pitch rates, boosted by vectored thrust, are up to twice as fast as the F-16. In fact, the F-22's pitch rate is so fast that it is inhibited by a soft stop in the aft movement of the sidestick. Pulling the stick through the stop overrides a limit in pitch acceleration. The stop is there to remind the pilot that the F-22 is about to respond fast, and the BRAGG valve will respond in turn. Lockheed Martin engineers and pilots have named the maximum pitch-rate regime "Mongo mode" in tribute to the horse-punching heavy from the movie *Blazing Saddles*.

A very important point is that the large flight

envelope is usable in combat, because the systems have been fully tested to high alpha (the fuel-system rig, for instance, goes to 60 degrees), and because the FCS protects the airplane against departure. At least some air-show maneuvers in recent years could not be performed without disabling FCS limiters, which normally constrain alpha and pitch rate. This may be acceptable for an expert pilot who is not worried about getting shot but would be as dangerous as the enemy in a combat.

The F-22 pilot who decides that the tactical situation warrants high-alpha, low-speed maneuvering may be reassured by the fighter's controllability and thrust-to-weight ratio. The F-22 should be able to end a maneuver rapidly when required and will accelerate quickly to a safer combat speed. "It will be a great air-show airplane, too," Metz adds.

Is stealth or agility more important? The answer may come during tests of the F-22, which, one program official says, "will be evaluated against real and simulated adversary aircraft." But the results may be secret; the Air Force does not like to talk about its collection of Russian aircraft any more than it likes to talk about their base, at Area 51 in Nevada.

The F-22 is claimed to have more than twice the range of the F-15C at subsonic speed, with a greater margin when the mission includes supersonic flight. Such numbers have to be treated with caution. In this case, the comparison is probably based on a full missile load and internal fuel only. The F-22's internal fuel load is greater than that of an F-15C with three 600-gallon tanks, and it has much less drag, so it should have a greater combat radius on a similar mission profile.

Designing so much capability into the aircraft has not been cheap, but it has been necessary, because of a characteristic of stealth designs. Since small details of the outer shape of the aircraft have a profound effect on RCS, it is not easy to modify a stealth aircraft to increase its range, improve its maneuverability, or add new weapons.

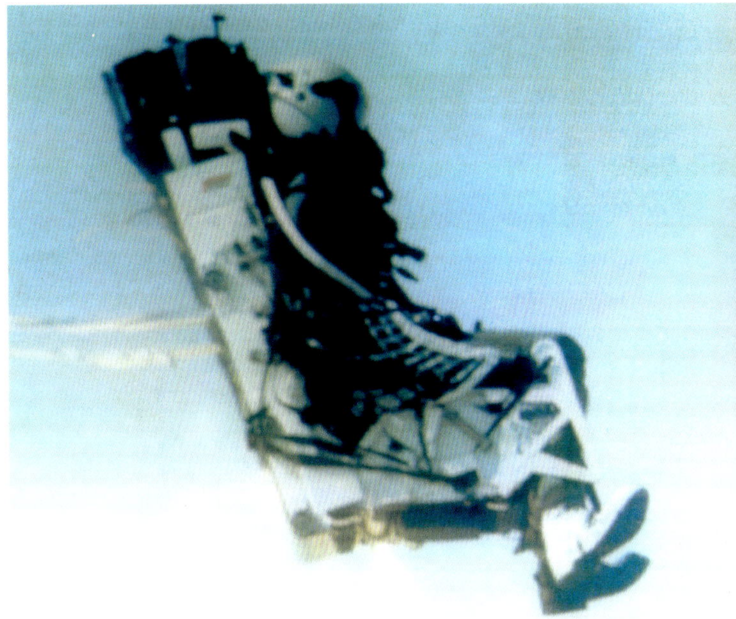

An advanced "anthropometric manikin"—a highly instrumented dummy—rides the F-22's modified Boeing ACES II ejection seat during a sled test at Holloman Air Force Base, New Mexico. The seat is modified with net-like arm restraints and an improved stabilizing drogue to cater for high ejection speeds. *U.S. Air Force*

In some ways, however, the biggest impact of stealth on fighter design has not been on the airframe or the engine, but on what may, in the 21st century, be the dominant feature of any military vehicle: the systems by which it gathers and processes information.

YF-22
N22YF

CHAPTER THREE

EVERY PILOT AN ACE

The war of August 2008 is 90 hours old. Sea and ground units will arrive, the planners calculate, about 36 hours after the last of the defenders on the ground chooses between surrender and annihilation.

Unless, that is, air power can slow the advance. F-15Es and F-16s are inbound to hit roads, bridges, tanks, trucks, and fuel dumps. One-on-one, they equal the enemy's modern fighters. But this isn't basketball. One-on-one is a win for the adversary, because it keeps Air Force fighters off the enemy's ground forces.

High above the attack aircraft, four F-22A fighters sweep into bandit country. Their mission is to engage the hostile fighters, prevent them from reaching the F-15s and F-16s, and destroy as many of them as possible.

The first YF-22 at low speed on its first flight. An early snag in flight-testing was that the landing gear refused to retract, a problem that Lockheed's flight-test manager, the late Dick Abrams, ascribed to "fascist software" in the vehicle-management system. Testing proceeded rapidly once a hard-wired switch was installed. *Lockheed Martin*

Gene Adam, McDonnell Douglas' cockpit-design guru, described air combat as "playing a three-dimensional video game, in constant fear of assassination, with the entire Los Angeles Rams front line sitting in your lap." The F-22 makes the job tougher. Because of speed alone, the video game will unroll two-thirds faster in front of the F-22 pilot.

Stealth gives the pilot a new dimension to consider. Like a submarine commander, the F-22 pilot must constantly weigh offense against concealment. Moreover, different radar systems can detect the F-22 at different distances, and radar can see it better from the side or the tail than head-on. Paradoxically, turning away from a radar can compromise the F-22's stealth.

Stealth presents designers with challenges. Blasting the sky with radar power is not an option. Remaining stealthy while acquiring useful information demands a new approach.

Paul Metz looks at the problem from a slightly different angle. "If you look at history, very few fighter pilots are effective," he says. In World War II, only 21 percent of fighter pilots made kills and

The second YF-22 resumed flying in October 1991 with an Air Force serial and was intended to support flight qualities and performance work during the EMD program. It performed 39 flights during this period, more than it had flown during Dem/Val, but was not considered critical to the program because of differences between the prototype and EMD designs. *Lockheed Martin*

about one in six of these (3.6 percent of the total) became aces. In Korea, the 4.8 percent of pilots who became aces made 38 percent of the total kills. "What if we can increase the ratio of pilots who make kills from one in five to one in two, or even one in three?" says Metz. The implications in terms of force effectiveness are clear.

Two particular aspects of the F-22's aerodynamic and systems design help accomplish that goal, by making the task of flying the aircraft as simple as possible. The FCS and carefree-abandon characteristics mean that the pilot does not have to watch G loading and alpha limits. In the systems, automation and self-test are the rule. Launching the F-22 is a matter of inserting a data transfer cartridge—which sets up the displays according to the pilot's preferences—switching the battery on, holding the auxiliary power switch in the on/start position, and setting the throttles to idle. The engines start automatically and the avionics run through their diagnostic routines, and within a classified but extremely short time, the fighter is ready to go.

But it is the F-22's sensors and displays that play the greatest part in making the pilot effective, which means resolving a fundamental conflict: automation is essential because no pilot can absorb all the data flowing into the system, but the pilot must be left in control of tactical decisions that only the human mind can make. The watchword in the F-22 cockpit design and the key to keeping the pilot in the loop without overloading the mind are to "maximize information and minimize data."

Basic concepts behind the avionics design include sensor fusion, combining data from all different sensors to display one target on the screen, and relieving the pilot of the need to monitor and compare different displays. Sensor management means that, in normal operation, the pilot does not control the radar, the passive electronic warfare system, or the data links. This is done automatically according to the tactical situation. "The pilot is a tactician, not a sensor operator," comments cockpit team leader Ken Thomas. Emission control (EMCON) is a task of the sensor management system, keeping tell-tale electronic emissions at the lowest possible level.

Until the enemy is within visual range, the pilot deals with this system through three full-color liquid-crystal displays that dominate the instrument panel. The central, 8-inch-square tactical situation display depicts the world around and in front of the F-22. The left and right 6-inch-square displays are assigned to defense and attack.

This developmental mock-up shows the main features of the F-22 cockpit: four primary displays, an up-front control for the head-up display, a sidestick, and a multifunction throttle quadrant. The large central screen shows the tactical situation. The right and left screens are respectively devoted to attack and defense, and present a subset of the same data in more detail. The lower screen is dedicated to aircraft systems. *Boeing*

Unlike today's cockpit displays, all three displays are in the same orientation—"God's-eye view," with the F-22's track pointed straight up the middle of the screen—and use identical symbols. (A fourth display is used for systems information.) "These [on the attack display] are the guys I want to kill," explains a Lockheed pilot, "and these are the guys who are trying to kill me."

The second YF-22, with YF119 engines. This aircraft was used for missile-firing trials—note the temporary, fixed spoiler in front of the ventral weapons bay and the missile symbols behind the inlet. The serrations on doors and other apertures are very apparent. Although the prototypes were not required to demonstrate low RCS, this task being performed by full-scale models, they did incorporate some low-RCS features and materials to show that they would work in flight. *Lockheed Martin*

Approaching the forward edge of the battle area (FEBA), the F-22 pilots accelerate through supersonic speed. The central display shows friendlies. Three blue circles mark the other F-22s in the formation. Four green circles represent bomb-laden F-15Es. Each circle has a vector projecting from it, its length roughly proportional to speed.

Even this display tells the pilots more than they know today, says Lockheed's Bo Meyer, a member of the F-22 team who spends weekends flying F-15s from the other side of Dobbins Air Force Base at Marietta, Georgia. On the F-15, he explains, all the returns on the radar screen look the same, regardless of speed or identity.

One basic principle, borrowed from commercial aircraft, is the "dark cockpit"; the absence of a light or a symbol means that everything is going smoothly. The displays are free from irrelevant data: The only ground-related symbol is a dashed line showing the FEBA.

The F-22s exchange their own positions via a secure intraflight datalink (IFDL) that also harmonizes the displays in the four cockpits, so that each pilot sees exactly the same situation. The IFDL operates at low power and in a radio frequency (RF) band that attenuates rapidly in the atmosphere, so it is difficult for an adversary to detect or track.

The F-22 formation leader's tactical situation display starts to look more ominous. Four yellow square symbols—unidentified targets—appear at the top right of the display.

Most of the display symbols differ in shape as well as color. This helps the pilot tell them apart instantly and makes them distinguishable if the pilot is wearing an antilaser visor, which blocks certain colors.

The side displays show detail that could not be added to the central display without making it cluttered. For example, SAM sites are identified by pentagons. On the defense display, the pentagons will become solid when the radars are operating, and a circle around each site shows its detection range against the F-22.

Sensor management and EMCON functions define a "globe" in the airspace around the F-22, divided into onion-like layers. In the outermost ring, targets are not yet close enough to be attacked. When a target enters the middle ring, the F-22 pilot can initiate an attack, but the target cannot. In the innermost zone, the two aircraft are committed to passing within range of each other.

In this case, the F-22 could identify the targets

All three tactical displays are in the same orientation and use the same symbols. Note the missile-range scale on the right-hand edge of the right (attack) display; there is also an altitude display on the right of the same screen. The lower (systems) display has automatically switched to show the number and location of missiles, gun rounds, and decoys on board. *Lockheed Martin*

with radar. But they are not close enough to present an immediate threat, so the computers inhibit the use of the radar.

Sensor fusion fills the gap. The F-22 has three sensors. In addition to the radar, it has an electronic warfare (EW) system that detects signals from other targets and a datalink, which imports information from Airborne Early Warning and Control (AWACS) and other aircraft. Today's fighters have all these sensors, but the F-22's passive, nonemitting devices—the EW system and datalink—are much more important.

The system architecture behind the displays is revolutionary. The radar, EW, and communications, navigation and identification systems are not stand-alone devices, but peripherals serving the GM-Hughes Common Integrated Processor (CIP). This CIP comprises two banks of 32-bit computer

The F-22's APG-77 active array radar is not the same thing as the electronically steered antennas used on (among other aircraft) the B-1, MiG-31, and Rafale, although both types are confusingly identified as phased arrays. An electronically steered radar has a single transmitter and receiver, and an antenna that comprises hundreds of "phase-shifter" modules that steer the beam. An active array consists of thousands of more sophisticated modules, each of them a tiny radar transmitter or receiver. *Northrop Grumman*

Missile launches were one of several aspects in which the YF-22 test program went further than the rival YF-23. The AIM-9 Sidewinder (top) is fired from a rail in the side bay, while the AIM-120 AMRAAM is ejected downward from the ventral bay before its motor ignites. The missiles were unarmed. *Lockheed Martin*

modules housed in the forward fuselage. The square, book-sized modules plug into two racks in the fuselage, automatically connecting to the power supply, the databus, and a liquid cooling circuit. Every function is backed up; if a module performing a critical function fails, its software will be loaded on to either a spare module or one that is doing a less important task.

On today's fighters, each sensor has its own display and it is up to the pilot to blend them into a single tactical picture. The F-22's computers sort the sensor data into "track files." For example: If an intercepted radar signal comes from the same location as a target tracked by AWACS, it will go into the same track file. Next, the computer picks the best data from the file: AWACS will have measured the target's range and speed, but the F-22's EW provides a more accurate bearing and can identify it. Lockheed engineers stress that sensor fusion is hard to do on any other aircraft, where each system has its own different computer.

As the distance closes, the symbols change to red triangles, indicating that the targets are hostile. The displays identify them as Sukhoi Su-35s.

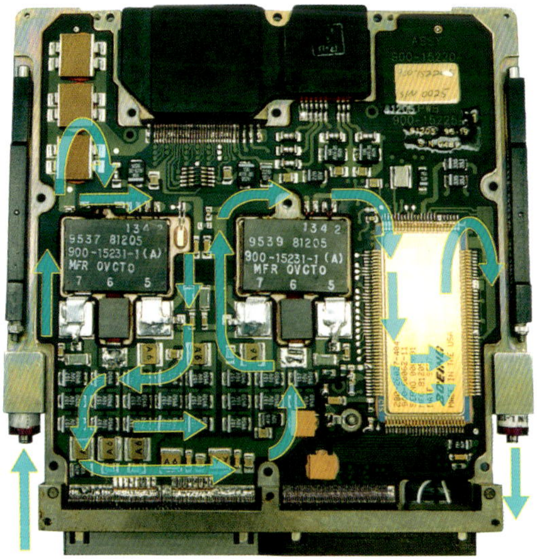

This power supply module is typical of the modules that make up the F-22's integrated processor. It measures 6.4 inches by 5.9 inches, is just over 0.5 inch thick, and weighs 1.8 pounds. When it is plugged into one of the F-22's built-in avionics racks, it is automatically connected to the databus, the power supply, and the liquid cooling circuit. *Boeing*

The United Kingdom-made HGU-86/P helmet, part of the Boeing-developed life support system, is seen here with its chemical and biological warfare mask and cape attached. The helmet is designed to remain attached even in high-speed ejections and incorporates a rear pressure bladder that pulls the oxygen mask tightly onto the head when positive-pressure breathing is in use. *Bill Sweetman*

In our simulated engagement, one of the targets has given the game away by using radar. The F-22's Lockheed-Sanders ALR-94 EW sensor suite "does not compare with anything out there today—it's vastly superior," remarks a Lockheed engineer. It can determine the target's bearing and, to some extent, its range. The ALR-94 can even be used to provide mid-course guidance for AMRAAMs—making a 'silent' AMRAAM attack a possibility

The CIP computes the range of the hostile's radar against the F-22 at its current bearing. It appears on the defense screen as a blue cone emanating from the target. The CIP will do the same for any SAM radars, placing a circle around them on the defensive display. If the F-22 turns to present its more reflective side or rear to the radar, the envelope will expand visibly. The pilot can choose whether to risk detection or change course.

The decision to engage is a no brainer: The Su-35s are headed for the friendly F-15Es, and the F-22s have the advantage of surprise. The leader taps a bar on the throttle. Immediately, each Su-35 is marked by a white

In-flight refueling was accomplished early in the YF-22 flight-test program and was used routinely to extend the length and productivity of each flight. The rotating in-flight refueling (IFR) receptacle is similar to that used on the F-22A. The prototypes were designed to represent the performance and handling of the final design, with similar thrust and loaded weights, but they were built quickly, sturdily, and somewhat overweight, and could not accommodate as much fuel as the production aircraft. *Lockheed Martin*

circle with a number, on both the main tactical display and the attack display.

The word SHOOT appears on the attack screen and the head-up display.

The purpose of the automated "shoot list" is to help the F-22 pilot make a good decision quickly, rather than making the pilot spend valuable seconds working out the best one. The computers prioritize the targets depending on their speed, position, and type.

The pilot can override the shoot list, which is one of a number of techniques pioneered by the U.S. Air Force's Pilot's Associate program. One of the goals of Pilot's Associate was "adaptive aiding" in which automation was there to help the pilot in high-workload situations but would not take over against the pilot's wishes.

Similarly, the defensive screen will show countermeasure and maneuver options against an imminent threat; the pilot can ignore them, evade manually, or consent to the automatic use of countermeasures.

The system now needs accurate target tracks, so the computers will allow more use of the radar. The Northrop Grumman/Raytheon APG-77 is the first fighter radar with an active-array antenna, comprising nearly 2,000 finger-sized transmit and receive modules embedded in a fixed array. The Air Force has pursued this technology since the early 1980s for three main reasons: power, agility, and reliability.

The F-22 radar has a peak power in the megawatt range. This allows it to put a great deal of power on the target in a very short space of time, gathering data before the target's electronic jamming or surveillance systems can respond. It makes it easier to use advanced modes to identify "noncooperating" targets at long range.

The agility of the beam—its ability to change in direction or waveform—is better than any other type of radar. In particular, the F-22 radar can switch among its 18-plus modes—so-called "interleaved" modes—so quickly that the modes appear concurrent to the avionics system.

Agility opens the way to the low probability of intercept (LPI) techniques, which the radar uses to perform its functions without betraying the fighter's own presence. Because the beam can scan instantly, it does not spray radiation across the sky as it sweeps from one target to another. The advanced solid-state power supply also responds quickly; an LPI radar never uses more power and never transmits longer than it has to.

Reliability is very important. The APG-77 has no antenna drives and no failure-prone electromechanical components such as rotary waveguide joints. The high-power transmitter/receiver and the single-channel amplifier and power supply are also gone. As for the transmit/receive modules, dozens of them can fail before the radar's performance is noticeably affected. Another feature of the active-array radar is that it is possible to add smaller arrays of modules, to extend its field of view. The F-22 designers planned to do this from the outset. The side arrays were deferred to save money early in the program, but space, cooling and software to support them are already in place. The USAF may well decide to reinstate them, giving the F-22 radar a horizontal field of view of al most 270 degrees.

This Kaiser Electronics Agile Eye helmet-mounted display is a forerunner to the HMD that will be used on the F-22. A tiny cathode-ray tube in the back of the helmet projects symbols and target data in front of the pilot's eye, via a combiner and a chain of mirrors. A sensor in the cockpit tracks magnetic inserts in the helmet, so the system knows where the pilot is looking. *Boeing*

On the side of the attack screen, a vertical bar shows the range to each target, and the maximum range of the F-22's AMRAAMs. Another symbol shows the range of the Su-35's missiles.

Dave Ferguson takes off from Palmdale, California, for the first flight of the YF-22 on September 29, 1990. This aircraft, bearing the civil registration N22YF because it was being operated by Lockheed, had GE YF120 variable-cycle engines, and engine delivery slightly delayed the first flight. The initial sortie was a short hop from Palmdale to Edwards Air Force Base in California, where the test program took place. *Lockheed Martin*

Once the targets have been selected, the missile-range bar gives a clear indication of what the F-22 and its missiles can do against these targets at this speed. As radar-guided missiles have become more reliable and more lethal, beyond-visual-range tactics have become more complex. A pilot under attack may dive sharply, hoping that the missile will slow down in the thicker air at low altitude and run out of energy. Firing early allows the attacker to break away. Firing later means that the missile will have more energy when it reaches the target and will be harder to evade.

While the F-22 pilot is working on attack options, the tactical and defensive displays continue to provide a rapidly assimilated picture of the larger battle and potential threats.

The leader and wingman fire two AMRAAMs each. On the missile-range scale, a small bar marks the point at which the AMRAAMs' own radar seekers will lock-on. As the fourth missile "goes autonomous," the F-22s turn sharply away.

The Su-35 pilots see nothing until the AMRAAMs lock them up. Warning systems activate with a blaze of lights and a chorus of alarms. The aircraft shudder as decoy launchers fire. The second-element wingman has two seconds more warning than the others and breaks right. His AMRAAM follows one of his radar decoys. The leader's wingman ejects, thinking his aircraft is hit. Moments later, he is right. Hanging in his parachute, he counts three smoke trails winding toward the cloud deck. No other 'chutes.

Air Force Brigadier General James Fain, F-22

program manager during Dem/Val, once remarked that "the idea is to be back in the bar drinking a beer before the other guy knows what hit him."

New targets have appeared on the screen and used their radars. They are FC-1s, agile lightweight fighters built in China.

The defensive display shows that the direct course to the FC-1s lies across an active SAM site. The F-22 leader decides to take an indirect route, which should stay out of the SAM's lethal envelope. As the aircraft turn side-on to the radar they sacrifice some of their stealth. On the defensive display, the circle that symbolizes the SAM's radar coverage expands correspondingly and, for a few moments, covers the F-22s.

The F-22s are moving away from the SAM site and it cannot shoot them, but a warning is passed to the FC-1s and the fighters accelerate to meet the attack.

The F-22 is designed to avoid the visual-range battle by stealth (see the enemy first) and by speed (giving the F-22 pilot the option to give or refuse combat). "We don't want to mix it up in a knife fight," remarks a Lockheed pilot, "but it is going to happen, and we'll be able to handle it."

In a visual-range combat, much of the symbology on the attack display will be repeated on the wide-angle head-up display. Also, after years of indecision and spurred by Russian developments, the U.S. services are developing a helmet-mounted display that will keep pilots supplied with target data wherever they look. The F-22 will enter service with the Joint Helmet-Mounted Cueing System (JHMCS) and AIM-9X missile for off-boresight engagements. (JHMCS is being developed by Vision Systems International, an Elbit/Kaiser joint venture.)

Before anyone can take a long-range shot, the eight fighters are embroiled in a "furball" trailing over miles of sky. The FC-1 pilots should have a "first-look" advantage, with their smaller airplanes, but the U.S. pilots unaccountably seem to be able to see better. Controllers notice that the F-22 pilots are using voice radio less than the FC-1s. Are they telepathic as well?

The first Boeing 757 is being modified as a flying avionics laboratory, with the F-22's radar and a "wing" above the cockpit to accommodate the F-22's large wing-mounted electronic warfare and communications antennas. The interior will include an F-22 cockpit mockup and workstations for test engineers. *Boeing*

An FC-1 positions on the lead F-22, closing from the rear quarter. As the pilot's thumb tenses on the fire button, the F-22 stands on its tail, executes a rapid roll and fires a missile in his face, almost directly backward along its flight path.

"It's remarkable how much less you talk when the datalink is there," remarks a Lockheed pilot. "The first thing you learn is that the amount of information that you share is tremendous." This is why the F-22s seem so coordinated: Even if only one F-22 is tracking an adversary on radar,

Some key similarities and differences between the Dem/Val and EMD F-22 designs are apparent here. The YF-22's cockpit is further aft and the inlets encroach on the pilot's view. The outer ailerons are small and tapered, and the wing and tail shapes are different. The YF-22 also has a dorsal speed brake. One shared feature, the "cat's-eye" sculpting of the control-surface junctions, is very clear in this photo. It avoids a non-stealth 90-degree step that would otherwise appear when the control surfaces moved. *Lockheed Martin*

the other three pilots can see it on their displays. "So if my buddy here traps someone on his six, I can see it happen and save his bacon once again," the pilot adds.

The datalink and the helmet-mounted display account for the F-22 pilots' apparently superhuman eyesight and the speed of the leader's defensive shot. Even if you know roughly where to look, it can take seconds to find another aircraft in the sky. But the helmet display includes an arrow that guides your eyes to the target and then places a cueing symbol over it.

Windows for the electro-optical Missile Launch Detection (MLD) system are located around the forward fuselage. The MLD system picks up the distinctive infrared signature of a missile's rocket plume and warns the pilot that a missile may be coming from a certain direction. Four dispensers for flare, chaff, and active radar decoy cartridges are installed in the lower wing surfaces.

Another technology that is likely to be used on the F-22 is three-dimensional sound, using a digital processor to create a voice warning that appears to come from the exact direction of the threat, or to make the wingman's voice come from the wingman's position.

The F-22 display system has been extensively simulated since the late 1980s, including many real-time sorties using multiple interlinked dome displays. The results, says Lockheed Martin, show that the F-22 system is intuitive and easily learned and raises the performance level of an inexperienced pilot.

Demonstrating such concepts in a simulator and making them work in a real aircraft are two different things, and it is a measure of the avionics development task that six of the nine EMD aircraft, a modified Boeing 757 transport, and an extensive ground facility in Seattle are dedicated to avionics testing.

The Block 2 configuration adds radar modes and some EW functions and should be available in mid-1999. Block 3, originally planned as the final pre-IOC release of the software, should be released in April 2000 and includes all EW functions. It will be followed in late 2000 by Block 3.1, which includes provision for JDAM.

By late 1997, every piece of hardware intended for the system was built and working in the laboratory, including a complete radar array. Another encouraging sign: the 1.7 million software lines of code in the current system are, by computer standards, relatively close to the 1.3 million that was predicted in December 1990.

One potential snag has arisen from the huge mismatch between the pace of the F-22 program and the progress of computing technology: components that were state-of-the-art when the fighter was designed may be obsolete, and difficult, expensive, or impossible to acquire, by the time production starts. A special IPT team within the avionics program is responsible for identifying such components and developing substitutes, using commercial technology to the greatest possible extent.

The F-22 leader gives the bug-out call. Crossing the FEBA, the F-22 leader begins to wind down from the immense tension of the last 20 minutes. The tactical display shows four F-15Es heading back from their targets—no losses.

The F-22 leader looks forward to a shower and a few hours of rest. Besides, that last hard break seems to have tied her bra into some kind of a knot.

CHAPTER FOUR

NEW WORLD ENFORCER

The Raptor's first flight on September 7, 1997, marked only the midpoint between the start of EMD and the fighter's entry into service. Nine EMD aircraft are being built (two of the eleven aircraft in the original contract were eliminated at the beginning of 1993). The first three (4001–4003) are dedicated to airframe and engine testing and weapon release clearances. The second of these is due to fly in April 1998, around the same time as the first aircraft is delivered to Edwards Air Force Base in California on a C-5 transport, and the third is due to fly in September. They will have nonstandard displays, no mission avionics, and simpler, flight-test-dedicated communications equipment. The fourth through ninth aircraft (4004–4009) will all fly between April 1999 and May 2000, and all are dedicated to avionics testing. Flight-testing will therefore get under way in earnest in mid-to-late 1998, when two or three aircraft are available. The entire program will encompass more than 2,400 flights and 4,350 hours and will continue until 2003.

"Planform alignment" is a key principle of stealth technology. Radar reflections from the edges of the airplane can be reduced, but not eliminated. To reduce the probability of detection due to these residual reflections, the main plan-view edges—wings, tail surfaces and inlet lips—conform to a few alignments. Doors and other apertures are angled or serrated, with edges parallel to the wing and tail edges. The strongest reflections, at right angles to the edges, are concentrated in a few directions and are pointed away from the head-on aspect, so that they will not dwell for long on any receiver. *Lockheed Martin*

Low-rate production is due to start at the beginning of 1999; high-rate production will not be approved until virtually all the testing is finished, in late 2002, and the fighter will not become operational until late 2005.

Technical problems have little to do with this very deliberate pace. Development has been a success story. The engine passed its Critical Design Review in 1992, and the aircraft did the same in February 1995. Problems such as weight gains, RCS snags, and a shortfall in engine efficiency did occur, but they were discovered well before the first flight and were resolved without delaying the program.

September 7, 1997: On a warm southern Sunday morning, the first F-22A sits in a heat-haze generated by its idling F119 engines as pilot Paul Metz exercises the controls. Here, the rudders are deflected outward and the flaperons and ailerons are deflected upward, the configuration used for deceleration after touchdown. *Bill Sweetman*

Money has been a different matter. Since the EMD program started in August 1991, budget cuts moved the first flight from August 1995 to May 1997—a date that slipped to September without affecting the overall schedule due to a number of individually minor problems—and have delayed initial operating capability from 2001 to 2005. The Pentagon has reduced the planned F-22 fleet twice—from 648 to 442 aircraft, in the 1993 "bottom-up review" of U.S. defense plans, and then down to 339 in the mid-1997 Quadrennial Defense Review (QDR). This review also cut the peak production rate from 48 to 36 aircraft a year.

In July 1996, the Air Force deferred development of the F-22B two-seater to save money and eliminated two F-22Bs from the test program. This was not a painless decision, but the fighter's carefree handling and straightforward flying qualities should make it easy and safe to fly, while recording devices and the debriefing functions built into the Boeing-developed training system allow a pilot's performance to be reviewed on the ground.

The budget changes have been a tremendous problem. Whenever the program is slowed down, each IPT has to restructure its work so that it is finished at the right time, neither too late (which

The F-22A taxies out for its first flight. Originally set for the end of May 1997, the flight was repeatedly delayed by a series of minor problems, including a foreign object damage incident that damaged one of the F119 engines, a fuel leak, some electrical problems, and software difficulties. The first flight delay is not expected to have any impact on the overall program schedule, and Lockheed Martin program manager Tom Burbage says that the team has learned a great deal about the aircraft in the process. "We knew that the airplane was talking to us," he comments about the early delays, "but we didn't know what it was saying." *Bill Sweetman*

delays other teams) nor too early (which is wasteful because the team will be idle). The new schedules have to be compatible, so that one team is not left waiting for another team's work, and to ensure that teams are not conflicting in their use of test facilities.

The cuts waste hundreds of millions of dollars. The money that is cut in one year must all be spent in a later year to do the same work. Moreover, many costs are fixed—such as administration, security, and the maintenance of facilities—and the total spent increases as the program is stretched out. Lockheed Martin managers estimate that every dollar taken out of the program now means three dollars of extra cost to complete the effort.

The industry team has changed over time, as mergers have reshaped the industry. Of the seven companies that bid on ATF in 1986, only three survive at the time of writing, and one of those (Northrop) is due to be acquired by Lockheed Martin before this book appears. In March 1993,

Paul Metz rotates the F-22A Raptor for its first flight, at Dobbins Air Reserve Base in Marietta, Georgia, site of the final assembly plant and program headquarters. The first flight was made at Marietta because the program managers planned to fly the aircraft to Edwards Air Force Base rather than remove the wings and load it onto a C-5. However, plans have changed and the first aircraft will be transported to Edwards. *Lockheed Martin*

Lockheed acquired the Fort Worth, Texas, division of General Dynamics, increasing its share of the program from 35.0 percent to 67.5 percent. Two years later, Lockheed and Martin Marietta merged into Lockheed Martin. Partly in response, Boeing and McDonnell Douglas merged in August 1997. Boeing's share of the F-22 program is now part of its McDonnell Aircraft and Missile Systems business unit. Raytheon acquired key avionics supplier Texas Instruments and prepared to acquire GM-Hughes. In 1995 Northrop Grumman acquired radar integrator Westinghouse and is now set to become part of Lockheed Martin.

A review in late 1996 showed that the F-22's costs were likely to rise more than predicted, because defense industry costs are expected to rise faster than the government-wide inflation rate on which the Pentagon's budgets are based. The total EMD cost, including Lockheed Martin and Pratt & Whitney contracts, and work done by the Air Force, now seems likely to exceed $18 billion in FY1997 dollars, including the sums already spent or committed. The projected average flyaway price of the F-22 is now $70.9 million in 1997 dollars.

The long development process has given the F-22's critics time to marshal their arguments against the fighter. The gist of the case against the F-22 is that it is a Cold War weapon, designed to fight an enemy that no longer exists. The Pentagon could save money by improving the F-15 and F-16, critics argue, and buying the smaller, less costly Joint Strike Fighter.

The Air Force's defense of the F-22 is far-reach-

After starting the aircraft—with a burst of smoke from the auxiliary power unit—Metz took off with 100 percent military power and climbed at 250 knots to 15,000 feet, where he performed a series of flying-qualities maneuvers at pitch angles up to 14 degrees. The power on climb out—without augmentors—was also impressive: the F-22 took off at about 145 knots, and climbed to 15,000 feet with the landing gear extended, and Metz had to set a steeper-than-expected climb angle to keep the fighter below its maximum gear-down speed. *Bill Sweetman*

ing. In the latest revision of the Air Force's post-Soviet doctrine, air and space superiority is listed as the first of the Air Force's "core competencies." Air and space superiority is intended to provide U.S. forces with freedom of action, while preventing hostile aircraft and missiles from interfering with U.S. operations and denying them sanctuaries where they can operate.

"Too many people fail to understand how the country depends on air dominance," Air Combat Command chief General Richard Hawley remarked in January 1997. Hawley pointed to Rivet Joint and Joint STARS, two systems that provide vital information on hostile air and ground forces and are carried on modified C-135 and 707 transports. "How long will that information be available if those aircraft are threatened by long-range AAMs launched from sanctuaries protected by surface-to-air missiles? Will we be able to sustain precision attack operations against adversary fighters? Will ground forces be able to maneuver as they did in Desert Storm if the enemy's reconnaissance aircraft can see them?"

Desert Storm was a full-scale test of Air Force

The F-22 at low speed on its first flight. The F-22 proved to handle very well in formation flying, said Metz. "It's very agile, crisp, and precise." The landing gear was extended after a series of maneuvers, and Metz brought the F-22 down to 10,000 feet for a simulated approach. *Lockheed Martin*

Another proposed alternative to the F-22 is the Joint Strike Fighter, a multiservice, multimission aircraft designed around a single engine derived from the F119. It will be built in three versions—a conventional U.S. Air Force fighter, a carrier-based stealth strike aircraft for the U.S. Navy, and a short-takeoff, vertical landing model for the Marines. Lockheed Martin's X-35 design, with a clipped-delta wing and four tails, even looks rather like an F-22. *Lockheed Martin*

doctrine. Refined by tacticians such as Colonel John Warden, whose book *The Air Campaign* was the planners' bible, the Air Force doctrine that underlay Gulf War plans placed more emphasis than ever on air superiority. Most of the opening sorties of the campaign were aimed at the Iraqi air force and air defense system.

The Iraqi air force was paralyzed. Only on the second day did the Iraqi air force fly more than half as many sorties per day as it had done, on average, on January 1 to 15. On about the ninth day of the war, sorties against coalition aircraft ceased, and a day later the Iraqi air force began to withdraw to Iran.

There were only four days on which Iraqi

Structurally, the F-22 is almost tailless: Little of the F-22's mass lies behind the line of the trailing edge. The horizontal tails are carried on booms projecting aft of the engine nozzles, and their root leading-edges fit into cut-outs in the flaperons: a unique layout that reflects the need to keep the root chord long. *Lockheed Martin*

aircraft—including support and combat types—flew more than 40 sorties. Not surprisingly, the air war was less than intense. About half the Iraqi fixed-wing aircraft shot down (16 out of 35) were destroyed in the first three days; most of the rest were bounced by Air Force F-15s while attempting to escape to Iran.

With air supremacy secured, coalition strike aircraft were able to hit targets throughout Iraq with minimal losses. B-52s and even C-130s took part in the bombing campaign. Friendly reconnaissance aircraft operated at will, and Iraq had no ability to track even the largest allied troop movements. The way was open to a ground campaign that was faster and incurred fewer casualties than most pundits had believed possible.

F-15s have shot down 96 adversaries with zero losses in air combat, including 33 out of 35 aircraft shot down in the Gulf. The Air Force argues, however, that a more lethal and survivable replacement is needed to counter the proliferation of advanced fighters and SAM systems. Two factors support the need for a fighter that outclasses the threat, rather than matching it (as an improved F-15 could). First, U.S. and allied forces in-theater are likely to be outnumbered in the early stages of a conflict as they arrive and

establish their bases. Second, the U.S. public and political leaders expect quick success and minimal losses.

The Soviet threat, as we knew it, has vanished, but future adversaries will use much the same technology that the intelligence community projected for the Soviet Union when the F-22 was conceived, because the best SAMs and fighters developed for the former Soviet Union are now openly sold on the export market.

In September 1996, in the clear, cool late-summer skies over Surrey, England, Russia's Yevgeny Frolov stole the Farnborough air show in the Sukhoi Su-37 fighter. Standing on its tail, slowing almost to a standstill, the fighter demonstrated that it can flip right, left, or backward to fire a missile in almost any direction. The Su-37's key features, including vectored thrust and foreplanes, will be used on the closely related Su-30MKI fighter, which Sukhoi is already building for India.

As noted earlier, the value of the Su-37's low-airspeed performance is controversial. But the question as to whether the F-22 will outperform the Su-37 in a "1-v-1" fight is far from the whole story. Air warfare is never 1-v-1, and future threats are likely to be different both in numbers and in type from what the United States fields against them.

Thousands of new and advanced fighters will

The F-22 cleaned up, on its first flight. The F-22 climbed to 20,000 feet before demonstrating a series of slow and fast throttle movements from idle to military power. It is easy to overlook an important fact about the airplane: This is how it will look when it goes to war, free of external tanks and missiles. *Lockheed Martin*

Metz taxies toward the parking area after the 58-minute first flight. Canopy and helmet are clues to the F-22's substantial bulk: The cockpit floor is not much lower than the forebody chine line. The F-22 has a new camouflage scheme, an overall gray with soft-edged darker areas on the wings, body, and tail. The base color is intended to match the luminance of the sky at typical combat altitudes and extreme visual range, while the darker patches send mixed signals to the eye, or to an electro-optical seeker with an edge-recognition algorithm. *Bill Sweetman*

The F-22 is highly blended: One-third of the total span lies between the wing attachment fittings. The inlets are widely separated and the ducts snake upward and inward, concealing the engines from radar. The inlets are fixed-geometry, one of many ways in which the Air Force decision to forgo a high-Mach capability (seldom used on the F-15) saved time, weight, and money. Boundary-layer turbulence is controlled by drawing air through pores in the duct wall, and the air is dumped overboard through exhaust grilles and a bleed door. *Bill Sweetman*

be built and sold in the next two decades, not just Su-37s and MiGs. France has exported its capable Mirage 2000 and is hoping to sell the excellent Rafale. Sweden and the United Kingdom have joined forces to market the small, cheap-to-operate Gripen, and the European partners behind the Eurofighter are striving for export sales.

None of these countries would knowingly sell aircraft to nations hostile to the United States, but fighters last longer than promises of eternal friendship. Iraq flew Mirages in the Gulf War. The British fought against U.S.-supplied A-4s and Israeli Dagger fighters in the Falklands.

U.S. government embargoes have generated new suppliers in the fighter market. In the aftermath of Tianamen Square, the U.S. government pulled out of the U.S.-Chinese Super-7 fighter project. China responded by developing the Chengdu FC-1, a fighter closely resembling the Northrop F-20 Tigershark, and Pakistan joined the project when the United States refused to deliver F-16s. The partners intend to build the FC-1 with a license-built Russian engine and with a choice of Chinese or Western avionics. The aircraft promises to be remarkably cheap.

Neither do fighters have to be advanced in all

The F-22 sits low to the ground, compared with most fighters, because there is no centerline pylon: The lower fuselage is occupied by weapon bays. This keeps the landing gear short and light (the main units retract outward with the wheels lying in the wing-body junction) and places most important systems chest-high for maintenance. *Bill Sweetman*

The nose probe, alpha sensors, and upper and lower blade antennas on the first F-22 are there for flight-test purposes only. The F-22 could be described as bristling with antennas if any of them were visible. The 30-plus apertures for the CNI and EW systems are all flush with the surface of the aircraft, including large arrays in the wing leading edges. The EW system includes azimuth and elevation antennas to provide three-dimensional target data. The rhomboidal patches on the side of the nose are the pressure ports for the air data system. *Bill Sweetman*

The F-22 returns to its parking point after the first flight. The wing-to-body blending and complex wing camber are very apparent in this view. The F-22 flew with an FCS software package that is intended to cover the entire flight envelope and all configurations; no further "releases" are planned. The thrust vectoring nozzles of the F119 engines are fully operational and could be seen deflecting as Metz tested the controls before the flight. *Bill Sweetman*

respects to pose a threat. There is a continuing market in fighter upgrades, in which new weapons and radars are grafted onto existing aircraft. In a 1-v-1 contest, a MiG-21 armed with modern Russian missiles is no match for an F-22; but it cannot be ignored as a threat in a real, more chaotic battle.

Where the F-22 should prove most useful is in its ability to give its pilot control of the engagement, however many adversaries may be involved. The F-22 pilot will have the speed and lethality to engage when conditions are favorable, the stealth and agility to decline combat when they are not, and the situational awareness to tell the difference.

Air Combat Command leader General Richard Hawley (left) congratulates an ebullient Paul Metz after the first flight. Metz described the new fighter as handling like an F-15, "one of several aircraft that are benchmarks for ease of flying. It's subconscious, like driving a car. The F-22 is very like an F-15 in the way that it lands and takes off. If you can fly a Cessna 150, you can fly this airplane." Metz will not talk about the most interesting comparison, however—between the F-22 and the Northrop/McDonnell Douglas YF-23, which Metz flew as chief test pilot for Northrop—saying only, "The Air Force made the right decision." *Bill Sweetman*

A fundamental and often overlooked point is that the need for the F-22 is not exclusively driven by the fighter threat and never has been. Many of its unique features—all-around stealth, supercruise, and high altitude—are equally important in countering SAMs.

SAMs cost less than fighters, and their operators need less training. They can be used to protect ground targets or to create sanctuaries for aircraft. Most of the coalition aircraft lost in Desert Storm were hit by SAMs, and SAMs protected many Iraqi targets against most aircraft.

Russia is selling its latest and best systems on the world market, ranging from the highly mobile Buk-M1 (SA-11) to the monster S-300V (SA-12). The S-300V system has two missiles, the smaller of which has a peak velocity of almost Mach 6 and is specifically designed to engage targets evading at 8 g. The larger missile attains Mach 8 and has an effective range of 60 miles. The Buk-M1 is the replacement for the Kub (SA-6) SAM that shot down Captain Scott O'Grady's F-16 over Bosnia.

A single system, mounted on 11 vehicles, has 36 missiles ready to fire at any time.

Stealth has been proven to be the most effective defense against SAMs. The F-22 can cruise on the edge of the Buk-M1's altitude envelope, reducing the missile's effective range; a larger system such as the S-300V is still a threat, but closes a much smaller area to a stealthy, high-performance aircraft.

There is no doubt that the threat is real and will continue to grow. Some F-22 critics acknowledge this, but suggest that there are less costly alternatives. One approach that has been proposed is for the Air Force to field an improved version of the F-15, while Lockheed Martin itself has studied some radically improved F-16s, including a delta-winged, stretched version with a wing not unlike that of the F-22.

One alternative to the F-22 is an upgraded F-15. McDonnell Douglas discussed this design in 1994, with a larger and much thicker wing which boosted the fighter's internal fuel capacity by more than 40 percent. Not shown here are other possible modifications, such as fixed-geometry low-observable inlets, that could reduce the F-15's radar signature. *McDonnell Douglas*

The snag is that such radical modifications are

This formidable array of ordnance and electronics is only part of the S-300V SAM system. It includes two missiles based on a common upper stage with either a large or small booster. The entire system is mobile and can be employed or disemployed in five minutes. Developed in the 1980s, the S-300V is ready for export and has been widely exhibited at trade shows. *Rosvoorouzhenie*

If the U.S. Air Force was to choose the upgrade path, Lockheed Martin would undoubtedly offer a radically improved F-16 such as this delta-winged aircraft, proposed to the United Arab Emirates in 1994–1995. Reminiscent of Fort Worth's ATF design, it would have 80 percent more internal fuel than an F-16 and would carry weapons more efficiently. *Bill Sweetman*

not cheap. An F-15 or F-16 with a new wing or engine would cost well over $2 billion to develop. This might not be as much as the cost of completing the F-22 program, but it is not vastly cheaper, and the resulting aircraft would not be stealthy, a supercruiser, or have the F-22's avionics. Neither would an improved F-15 be much less costly to buy than the F-22.

Alternatively, the Air Force could wait for the Joint Strike Fighter—a multirole aircraft that Boeing and Lockheed Martin are competing to produce for the Air Force, Navy, Marines, and export customers. From a distance, Lockheed Martin's Joint Strike Fighter design even looks like the F-22.

There are several good reasons why this is a bad idea. The Joint Strike Fighter is at the same stage that the ATF reached in 1987. It will cost more and appear later than advertised.

The Joint Strike Fighter would be more costly if the F-22 were canceled. The fighter's engine, stealth design, and sensor fusion philosophy are based on F-22 experience and will be more expensive and riskier without the lessons that Boeing, Lockheed Martin, and Pratt & Whitney will learn during the rest of the F-22 program.

Above all, the Joint Strike Fighter emerged from a tough process in which the services traded off their requirements and desires to arrive at a single, affordable aircraft. One of the basic assumptions was that the Joint Strike Fighter would never fly against a first-rank threat without the support of Air Force or Navy combat air patrols. Because of this, the Joint Strike Fighter's air-combat attributes were not stressed. It will not supercruise, it will not be very fast in transonic acceleration, and its sustained G levels will be lower than an F-16.

So far, the F-22 has survived. The Pentagon's 1997 review neither canceled the program nor cut it back to a token "silver bullet" force—two options that were moot before the review was published. Combined with a supportive Congress, this goes a long way toward stabilizing the program for the life of the present administration, and EMD will be almost complete by the time the

Congressional worries about a "tactical aircraft train wreck" peaked in early 1997, with increasing rates of spending on the F-22, Joint Strike Fighter, and this aircraft, the Navy's F/A-18E Super Hornet. Some Washington figures have even mentioned the Super Hornet as a substitute for the Raptor—although it is similar in size and inferior in most aspects of performance to the F-15. *Boeing*

The face is familiar: Boeing's X-32 Joint Strike Fighter contender recalls Seattle's ATF design in the shape of the inlet and forward fuselage. It features a thick, one-piece delta wing, however. Boeing and Lockheed Martin are each building two Joint Strike Fighter prototypes, due to fly in late 1998. Potentially, the U.S. services could need 3,000-plus Joint Strike Fighters. *Boeing*

next president is inaugurated.

Even the cut in production was not as bad as it seemed, because the QDR encouraged the Air Force to designate the F-22 as the ultimate replacement for the F-117 and F-15E. Already, the Air Force is talking about a requirement for 200-plus strike-modified Raptors to be delivered starting in 2012. It is early to talk about specific features, but the new version will probably carry new weapons, and might have some weapon bay modifications. New sensors are a possibility, perhaps including the introduction of a thermal imaging and infrared search-and-track (IRST) system. IRST was part of the original ATF requirement but was deleted during Dem/Val after tests showed that its potential did not justify its weight and cost. Nevertheless, the Air Force and Lockheed Martin have continued to develop IRST, and a low-observable IRST window for the F-22.

Exports are a possibility. Pentagon policy precludes final contracts until initial operational test and evaluation is complete, in 2001–2002, but that does not prevent Lockheed Martin from briefing export customers. Potential customers include F-15 operators such as Israel, Saudi Arabia, and Japan. South Korea is considering a high-end fighter to complement the F-16, and Lockheed Martin is looking at the possibility of selling small "silver bullet" F-22 fleets to operators of modern but nonstealthy fighters.

The F-22 has no real competition. The most comparable aircraft outside the United States might be the Mikoyan 1.42—another large, stealthy, highly supersonic fighter—but development of this aircraft has stopped. Currently, the closest competitors are probably the large, powerful, and very nonstealthy Su-37 and the Dassault Rafale. The latter is a capable aircraft, but designed to a different set of requirements: a multirole aircraft, being produced in similar versions for the French Air Force and Navy, Rafale displays a unique approach to stealth using a combination of passive and active measures, terrain avoidance, and stand-off weapons.

Another of Lockheed Martin's potential competitors, British Aerospace (BAe), has already fired the first salvo in a sales battle, claiming that the Eurofighter has 91 percent of the F-22's capability for less than half the cost. However, these statements are highly misleading. Eurofighter's "effectiveness" comparison is based on studies that looked only at a head-on radar/missile duel and, in any case, are based on BAe's estimates.

BAe's claims on cost appear to rest on a comparison between a flyaway cost (the price of one aircraft) and a unit program cost (the cost of the entire project, including research and development, logistics support, and many other items, divided by the number of aircraft built). Actually—and given that cost definitions are fraught with a lack of international consistency—the F-22's flyaway cost does not appear widely different from the $50 to $60 million figures quoted for Eurofighter and Rafale.

Fifty years ago, the North American XP-86 became the first all-new fighter to fly for the newly formed U.S. Air Force. The Sabre was the first fighter that truly belonged to the jet age. It had a reliable 4,000-pound axial-flow engine and a 35-degree swept wing, together with vital refinements such as powered controls, speedbrakes, and a radar/gyro gunsight. Compared with the straight-wing jets that preceded it, the XP-86 had a much larger combat-usable flight envelope: It could fight at 600 miles per hour rather than merely flying at 500 miles per hour, and it could dive at supersonic speed. It would have had no serious rival until 1955, had Britain's Labour government not supplied Rolls-Royce's best jet engines to the Soviet Union. Even so, it was substantially superior to any of its contemporaries, secured air superiority over Korea, and set the pattern for many later fighters.

Like the Sabre, the F-22 is so far ahead of its contemporaries that it is not fully understood. We can only hope that it does not take a war, this time, to get the message across.

APPENDIX

F-22 Specifications

Dimensions	
Wing span	44 ft 6 in (13.56 m)
Length overall	62 ft 1 in (18.92 m)
Height overall	16 ft 7 in (5.05 m)
Wing area	840 ft^2 (78 m^2)
Propulsion	
Powerplants	Two Pratt & Whitney F119-PW-100 augmented turbofans
Bypass ratio	0.2:1
Intermediate power	26,000 lb thrust (116 kN)
Augmented power	39,000 lb thrust (173 kN)

Weights	
Operational empty	31,700 lb (14,375 kg)
Internal fuel	25,000 lb (11,400 kg)
Clean take-off	60,000 lb (27,200 kg)
Max take-off	80,000 lb (36,300 kg)
Performance	
Maximum speed	Mach 1.8; 1030 kt (1900 km/h)
Supercruise speed	Mach 1.6; 920 kt (1700 km/h)
Radius of action	750 nm (1400 km)

F-22 Milestones and Events

November 1981	USAF identifies requirement for new air-superiority fighter
May 1983	RFP issued for ATF engine development
September 1983	Pratt & Whitney, GE selected to build ATF prototype engines. Concept definition contracts awarded to seven aircraft manufacturers
September 1985	Dem-Val RFP issued
March 1986	Navy agrees to study ATF as F-14D follow-on
May 1986	Plan to fly ATF prototypes announced
July-August 1986	ATF industrial teams formed: Lockheed, Boeing and GD versus Northrop and McDonnell Douglas
October 1986	Lockheed and Northrop announced as winners
July 1987	Lockheed abandons its original design
October 1987	New Lockheed design defined
August 1990	First flight of YF-23
September 29, 1990	First flight of YF-22 #1 with GE YF120 engines
October 30, 1990	First flight of YF-22 #2 with P&W YF119 engines
November 3, 1990	YF-22 supercruises at Mach 1.58
December 10-17, 1990	YF-22 conducts high-alpha testing
April 1991	Lockheed, P&W selected for EMD; production total reduced from 750 to 648 aircraft
August 1991	EMD contracts awarded
October 30, 1991	YF-22#2 returns to testing
Late 1991	Navy abandons NATF
December 1991	F-22 external design frozen
April 25, 1992	YF-22#2 damaged in accident at Edwards
June 1992	F119 passes critical design review (CDR)
January 1993	F-22 program "rephased" to save money. First flight delayed and EMD prototypes reduced from 11 to 9
March 1993	Lockheed acquires GD Fort Worth
December 1993	First F-22 metal is cut
February 1994	Production reduced to 442 aircraft in post-Cold War downsizing
September 1994	FY1996 F-22 funds are cut by 10 percent, further delaying the program
February 1995	The F-22 airframe CDR is completed
March 1995	Lockheed Corporation and Martin Marietta merge
Summer 1995	Wind tunnel tests completed
July 1996	Two-seat F-22 deferred
September 1996	First F-22 mid-fuselage arrives in Marietta
October 1996	Flight-test engines and aft fuselage delivered to Marietta
April 9, 1997	First F-22 rolled out; name "Raptor" officially announced
Summer 1997	Under the Pentagon's Quadrennial Defense Review, production total is reduced again, to 339 aircraft.
September 7, 1997	F-22 first flight
June 1998	First flight of second F-22
April 1999-May 2000	Remaining six EMD aircraft due to fly
Mid-1999	Air Force low-rate production decision
Late 2000	High-rate production decision
2003	Flight testing completed
2005	Initial operating capability

Level Flight Envelope

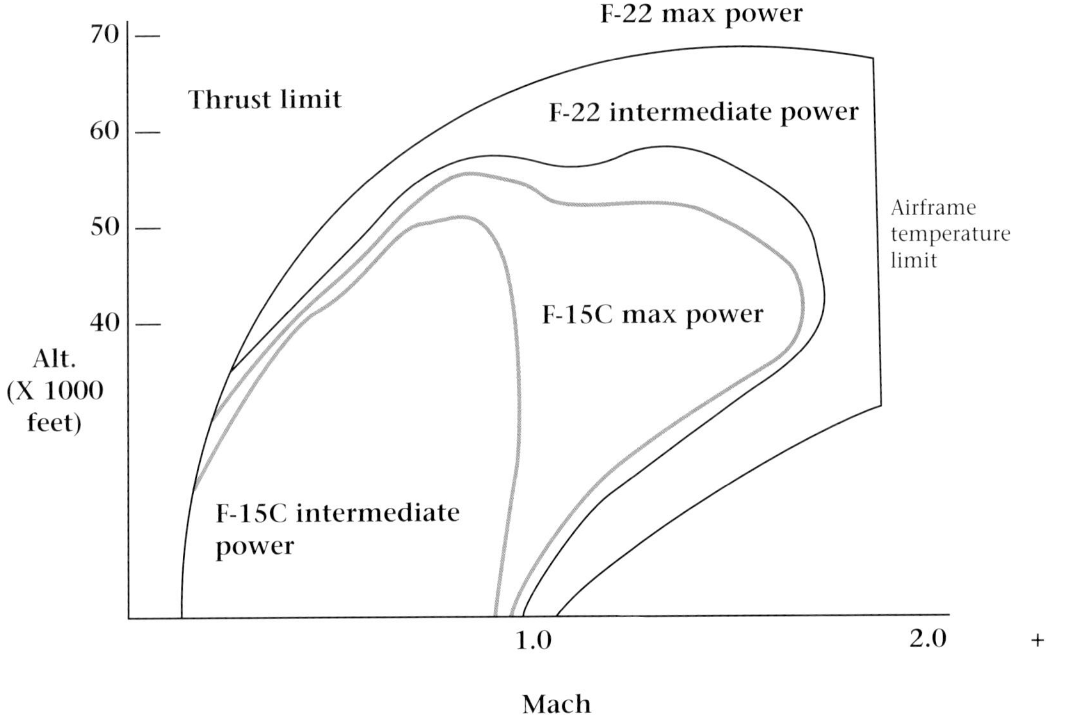

5g Maneuver Envelope

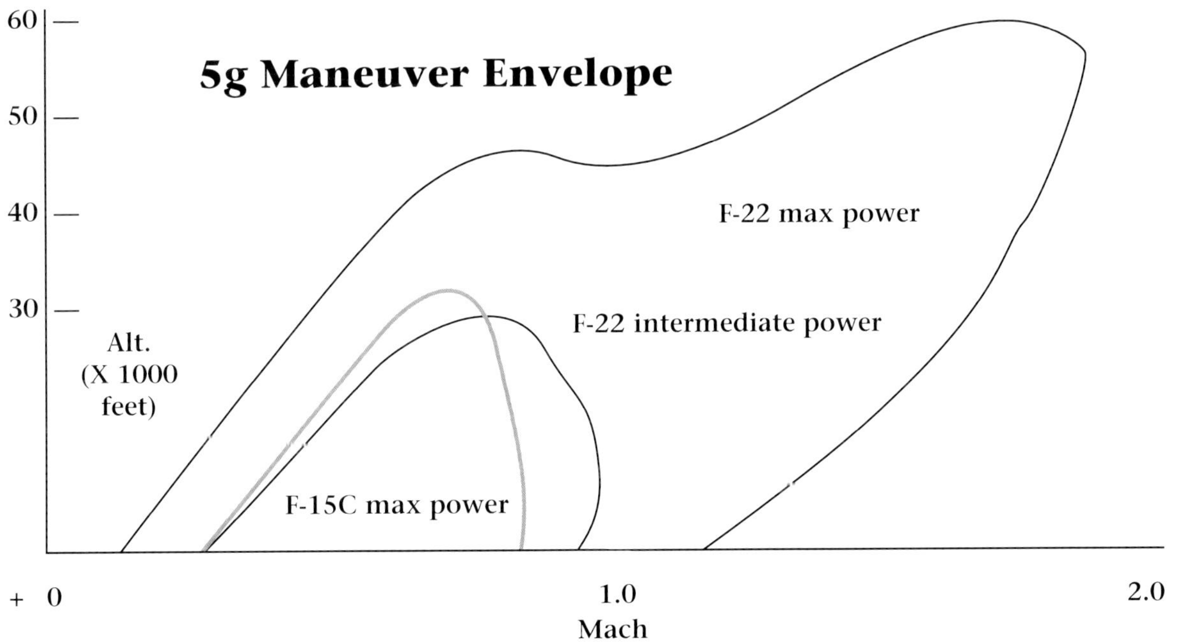

INDEX